CONTENTS

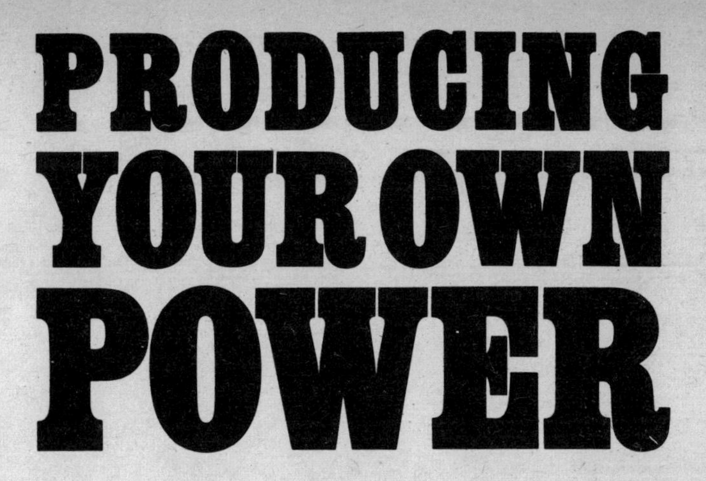

How to Make Nature's Energy Sources Work for You

Edited by
CARAL HUPPING STONER

Technical Consultants: Eugene and Sandra Fulton Eccli
co-editors of Alternative Sources of Energy
Illustrations: Erick Ingraham

Vintage Books
A Division of Random House
New York

VINTAGE BOOKS EDITION, *June* 1975

 Published in the United States by Random House, Inc., New York, and simultaneously in Canada by Random House of Canada Limited, Toronto. Originally published by Rodale Press, Inc., in 1974.

Library of Congress Cataloging in Publication Data
Stoner, Carol.
Producing your own power.
1. Power resources. 2. Power (Mechanics)
I. Title.
[TJ153.S795 1975] 621.4 74-23118
ISBN 0-394-71440-7

Manufactured in the United States of America

INTRODUCTION

Energy is all around us and manifests itself in many different forms—heat, light, sound, magnetism and gravity, movement, and all life functions. It's everywhere in great abundance. Since the beginning of time, nature has been producing and supplying us with energy in quantities so astronomical that we could never—not even with the most advanced technology—begin to tap and use it all. The force of moving water is so strong that about 3 million tons of rock are carried into the ocean by the earth's waterways every year, yet the world as a whole only taps 1 to 2 percent of this power. The World Meteorological Organization estimates that there is a wind power potential at favorable sites around the world of approximately 20 million megawatts. This is about 60 times the present generating capacity of electrical power plants in the United States. In this country alone, 3 billion tons of waste are accumulated each year, and the bulk of it is agricultural waste from which we could produce methane gas to replace a significant portion of our fuel needs. The sun is by far the largest source of energy. Solar energy falling on just the continental United States could supply us with 500 times our energy needs estimated for the year 2,000—if we knew how to harness it.

It is ironic that even with the abundance of energy with which nature endows us, we find this essential commodity in short supply. But actually, it's not energy that we lack, it's the know-how to tap more of this energy. Up until now we have learned how to produce appreciable energy from only one source: fossil fuels. And fossil fuels are far from the perfect source. By relying on them for virtually 99 percent of our fuel needs, we have been using only a minute part of all the energy available to us. A small fraction of 1 percent of the sun's energy is converted into plant tissue, and it is this tiny fraction that has produced all fossil fuels, in addition to all our food, all types of vegetation, and the myriad of products made from organic materials—like paper, petroleum products, plastics, etc.

And fossil fuels are environmentally destructive. Taking huge amounts of coal from underground and strip mines pollutes waterways and scars the land, leaving it barren and unsuitable for future use. Untold numbers of miners have become disabled or died because year after year they inhaled toxic fumes in coal mines. Oil spills from drilling and tankers off both the west and the east coasts have done great damage to shorelines and killed vast numbers of wildlife. Liquified natural gas is extremely explosive, and it has been the cause of many devastating fires.

Unsightly power plants, needed to generate electricity from coal and oil, and transmission lines needed to transport power from the plants to points of use, take up an incredible amount of space. The U.S. government estimates that by 1980 1 percent of our country's land will be covered with utility companies' equipment.

Fossil fuels, as we have been using them so far, are very inefficient. Five-sixths of the energy used in transportation, two-thirds of the fuel burned to generate electricity, and almost one-third of all the remaining energy is discarded as waste heat. That's *more than half* of the total energy we consume. Furthermore, a good part of these fuels that isn't converted into power or waste heat is emitted as polluting compounds like carbon and sulfur dioxide and nitrous oxide. The U.S. Office of Science and Technology has reported that 44 percent of the air pollution in this country today is caused by the emissions from motor vehicles.

Yet despite the many problems inherent in their use, we have become dependent upon fossil fuels over the decades because we have found ways to tap and utilize them easily and inexpensively. And, until recently, most of us assumed that they were in plentiful supply. Few, excepting some environmentalists and a handful of farsighted engineers and scientists, gave other energy sources much consideration because the world seemed to be endowed with an endless supply of fuel.

Now, after some dramatic brown-outs and black-outs, gasoline and heating oil shortages, and skyrocketing fuel prices, we know that coal, oil, and gas aren't going to last forever. As a matter of fact, they aren't going to last all that much longer at the rate we're using them. Over the last 50 years, the use of energy in all forms has been doubling every 15 years, with electrical energy (which consumes most of our fuel) doubling every 10 years. We're burning up in a few decades what it took nature millions of years to produce.

Decreasing our needs for such enormous amounts of power through conservation measures will certainly help, but it is crucial that we develop the technology that will enable us to utilize other forms of energy to meet our future requirements. In recent history, the only energy source aside from fossil fuels that we have given any serious consideration has been nuclear energy. Attracted by the fact that such tiny amounts of radioactive matter can yield such enormous quantities of energy, scientists have found ways to release energy by splitting nuclei. Since 1942, when Enrico Fermi set off the first nuclear chain reaction, nuclear power has been looked to as an answer to our energy problems. As of late, however, it looks as though the expanded use of nuclear fission may create more problems than it will solve. Nuclear power is just about limited to the generation of electricity, and although reactors require very little fuel, reserves of uranium are expected to last only until a few years into the 21st century. As they operate now, nuclear fission plants are less efficient in conversion than conventional fossil fuel plants.

But there are more fundamental problems with fission power. The safety of thousands or more large nuclear power plants is highly questionable. There are hazards involved with both mining uranium and disposing of radioactive wastes. Twice as much water is needed to cool nuclear plants than fossil fuel plants because of the intense heat they create during operation. Plant accidents that leak radioactivity into the atmosphere are not all that uncommon, and sabotage of these plants, although an improbability, could be disastrous.

Even work on new ways to extract coal and oil that were previously unavailable to us is not going to help solve any long-range energy shortages. If all mineable fossil resources were made available to us, we would still have energy problems. In a few centuries these will also be exhausted. And, if we deplete our supplies of these fuels, we would not be able to continue producing petrochemicals, plastics, medicines, and the other products that depend upon fossil fuels as raw materials.

Fortunately, we are now realizing that there are other huge reserves of energy that aren't many decades out of our reach. At long last, we are taking a long hard look at the other sources nature makes available to us every day—wind, water, organic matter, and the sun. And we're finding these sources to be very attractive. Unlike fossil and nuclear fuels, we can tap and use them while doing minimum damage to the environment. Solar, wind, and water power emit no solid or gaseous pollutants. Methane gas production actually rids our environment of wastes, since it converts farm wastes and garbage to fuel and valuable fertilizer. Methane, with an octane rating of 120, is a very clean fuel; it burns completely. Moreover, all these energy supplies are renewable and available in plentiful amounts in most parts of the world.

Although it seems that these so-called new or alternative energy sources have only recently been discovered, our use of them is not at all new. Throughout nearly all of history men and women depended upon these sources for all their energy needs. It was only some 150 years ago that we switched from renewable sources to the non-renewable sources that we're using today.

It's common knowledge that our first direct use of energy was made when primitive humans discovered fire, but it is interesting to note that windmills and water wheels were used both in ancient Persia and China. The Babylonians irrigated their fields with water pumped from windmills back in 1700 B.C. until the steam engine was invented in the 1700's, water wheels were the major source of power in this country. Before the Rural Electrification Program in the 1930's, wind generators were an integral part of rural American life. Millions were used to light farms and charge batteries. In the 17th and 18th centuries, European scientists were building solar furnaces and burning glasses to melt metals and other hard substances, and in the 17th century the first solar engine was invented. Early in this century, solar cookers, stills, and furnaces were in operation throughout the world. There were even a few solar-powered cars and boats in operation at that time.

A few years ago, we considered these alternative sources a part of our history; now we see them as playing a significant role in our future. The National Science Foundation/National Aeronautics and Space Administration's Solar Energy Panel, which did an investigation into the availability and utilization of alternative energy sources, believes that these sources have great potential. They concluded in their December 1972 report, *Solar Energy As a National Resource,* that by 2020 solar energy could economically supply 35 percent of our nation's heating and cooling load, 30 percent of our gaseous oil, 10 percent of our liquid fuel, and 20 percent of our electrical power. They also stated that we have had the technology needed to harness wind power for 60 years, and that it could be cost-competitive with electrical power from fossil fuels.

Hydropower already supplies one-fifth of the electricity produced in the United States, but this is less than one-third of the power we could tap from our rivers, streams, and lakes. U.S. waterways have a potential of supplying us with 80 million horsepower.

Methane gas digesters are also capable of producing a significant portion of our energy requirements. The NSF/NASA Solar Energy Panel believes that if we use our agricultural waste to fertilize algae, then use the algae to produce methane gas, we could significantly increase our gas supply by the early 1980's and produce as much as 100 percent of our gas needs by the year 2020.

Even wood, which today is generally considered a rather primitive and, in many cases inefficient utilization of energy, is thought by some to be a future source of energy. Some in the scientific community contend that 400 to 500 square miles of land could produce enough trees to keep a 1,000-megawatt power plant supplied with fuel at a cost substantially lower than if it were fueled with oil or coal.

Scientists and engineers have already, now in mid-1974, drawn proposals for municipal methane digester plants, one-and-two-mile-long solar farms, orbitting solar satellites, and huge, off-shore wind generator plants—all to supply electricity for thousands of homes. We have already begun to tap and utilize these alternative energy sources on a small scale, on farms and homesteads, residential homes, even office buildings.

In the past few years, many conferences have been held between the scientific community and industry in an attempt to show industry how it can apply alternative energy research to practical applications. And industry has responded. New companies are being formed to manufacture and market these energy installations, and much of this new hardware is already available in building supply catalogs. Plans for constructing solar-heated houses, modern wind generators, and bio-gas plants from readily available materials are emerging. Several systems have been built by people who no longer have to be dependent upon Con Ed and Exxon for their electricity and heating fuel. Magazines like *Alternative Sources of Energy, The Mother Earth News,* and *Popular Science,* are featuring articles on ways to harness energy at home.

Producing Your Own Power, one of the first books on small-scale power production, is more evidence of this alternative energy revival. It presents many of the recent developments in the field and illustrates how we, today, can begin to make nature's energy sources, which we have neglected for so long, work for us. The contributors are many of the people who have been creating all the excitement about "homemade" power. They are scientists, engineers, architects, educators, and homesteaders, most of whom have designed and built their own power systems. They come together here to discuss the limitations and potentials of small-scale power production and spell out many of the basics of harnessing energy from the sun, the wind, water, wood, and organic wastes.

C.H.S.

1 WIND POWER

Electric Power from the Wind

Henry Clews

Thirty or 40 years ago, wind–generated power played an active part in the electrification of many rural homesteads all across the country. Then the government pushed through the Rural Electrification program and within a few years banished these fledgling attempts at electrical self-sufficiency. The Power Company could provide more power at a lower cost by burning coal in huge plants and distributing the power on a network of wires to individual users.

But there was one important difference. The Power Company was consuming an irreplaceable natural resource to produce its power. The wind generators that were once essential power sources consume nothing. They don't give huge amounts of "cheap" power, but they do provide moderate amounts of free, non-polluting, environmentally safe power from an independent source which only costs you whatever you are willing to put into the apparatus for harnessing it. Back in the 1930's not too many people paid much attention to this difference. Today, however, things are changing. We here at Solar Wind have said "no thanks" to the Power Company and have installed our own wind generator. And other people are beginning to follow.

So much for the sermon. Now for a detailed account of the workings of a modern wind–powered generating system. Primarily, I will describe the operation of present–day production equipment which is available right now, ready-built and fully assembled. But for the many people who are interested in systems they might build at home on a limited budget, I have included a good deal of general information which, I hope, will prove of value.

Henry Clews atop the tower that supports one of his wind generators.

A complete, self-sufficient home wind–generating system consists of three or four parts: (1) the wind-driven power *plant* itself, (2) a *storage system* (usually a bank of batteries) to store the power for windless periods, (3) *conversion devices* to convert the generated or stored power to useable forms (usually from DC to 115 Volts AC or standard "house current"), and (4) an optional *back-up system* such as a gasoline or methane generator for times when the stored power is not sufficient to last through a long calm spell.

The Wind–Driven Power Plant

Virtually all electricity is produced by rotating electric generators which produce electricity by rotating magnets in front of each other. The Power Company uses huge generators turned by steam turbines (or in the rare case of hydroelectric power, by water turbines). The steam to turn the turbines is produced by boiling water over a coal fire (hiss, boo!). New atomic energy power plants function in the same way except that the heat to produce the steam comes from radioactive fission instead of coal. In an automobile, the generator is turned through a V-belt by the gasoline engine. Similarly, in a small portable power plant, the generator is turned by a gasoline, diesel, or L.P. gas engine. Thus all forms of useable electricity come from some type of a rotating generator which is driven by an external power source. The wind generator is no exception. A wind–driven generator consists of a rotating generator turned by a propeller which in turn is pushed around by the force of the wind upon it. The propeller can be thought of as a wind engine using wind as its only fuel.

Now, the amount of electricity that can be generated by a wind generator is dependent on four things: the amount of wind blowing on it, the diameter of the propeller, the size of the generator, and the efficiency of the whole system. Here are some specific examples to show you how this works. First consider an 8-foot diameter propeller with well designed blades having an efficiency of, say, 70 percent and a generator capable of delivering 1000 watts. In a 5 mph breeze you might get 10 watts of power from it; at 10 mph about 75 watts;

at 15 mph, 260 watts; and at 20 mph, 610 watts. As you can see, the more wind, the more power. But it is not a simple relationship. The actual power available from the wind is proportional to the *cube of the wind speed;* in other words, if you double the wind speed you will get eight times as much power.

Now let's consider a propeller with a 16-foot diameter and a similar efficiency to the first one. At a 5-mph wind we might get 40 watts output; at 10 mph, 300 watts; at 15 mph, 1040 watts; and at 20 mph, 2440 watts if the generator were capable of delivering this much power. As you can see, the power output of the 16-foot diameter windmill is about four times that of the 8-foot diameter windmill. This shows that the power is proportional to the *square of the diameter,* or that doubling the size of the propeller will increase the output by a factor of four. And there you have the two basic relationships which are fundamental in the design of any wind–driven power plant. A careful study of the table below will further serve to illustrate these relationships.

A view of the Clews homestead.

But what about efficiency and generator size? The efficiency (defined as the ratio of the power you *actually* get to the theoretical maximum power you *could* get at a certain wind speed) depends largely on what type of propeller you use. All modern electric wind-generating plants use two or three long slender aerodynamically shaped blades resembling an aircraft propeller. These efficient propellers operate at a high tip speed ratio which is the ratio of propeller tip speed to wind velocity. The Quirk's propeller, for example, runs at a tip speed ratio of about six while for some of the Swiss Elektro units this ratio runs as high as eight. This compares to ratios of one

TABLE 1. Windmill Power Output in Watts assuming 70% efficiency

PROPELLER DIAMETER IN FEET	WIND VELOCITY IN MPH					
	5	10	15	20	25	30
2	0.6	5	16	38	73	130
4	2	19	64	150	300	520
6	5	42	140	340	660	1150
8	10	75	260	610	1180	2020
10	15	120	400	950	1840	3180
12	21	170	540	1360	2660	4600
14	29	230	735	1850	3620	6250
16	40	300	1040	2440	4740	8150
18	51	375	1320	3060	6000	10350
20	60	475	1600	3600	7360	12760
22	73	580	1940	4350	8900	15420
24	86	685	2300	5180	10650	18380

to three for the slower–running multi-blade American water-pumping windmill. But while the latter type is less efficient, they do have a much higher starting torque, and their steadier speed at low wind velocities makes them more suited for pumping applications.

Ideally, the propeller of a wind machine used for generating electricity should have a cross-section resembling that of an aircraft wing, with a thick rounded leading edge tapering down to a sharp trailing edge. It should be noted, however, that the most efficient airfoils for aircraft propellers, helicopter blades, or fan blades (all designed to move air) are not the most efficient airfoils for windmills (which are intended to be *moved by* air). An old airplane propeller, in other words, has neither the proper contour nor angle of attack to satisfactorily extract energy from the wind. If you have the ability and time, and wish to construct a propeller of your own, you will find several good designs to copy in the *Proceedings of the United Nations Conference on New Sources of Energy.*[1]

The Generator

The generator itself forms the vital link between wind power and electrical power. Unfortunately, most generators which would seem to be suitable suffer from the requirement that they need to be driven at high speeds; they are built to be driven by gasoline engines at speeds from 1800 rpm to 5000 rpm. But windmill speed, especially in the larger sizes, seldom exceeds 300 rpm. This means that one must either find special low speed generators (which are expensive and cumbersome) or resort to some method of stepping up the speed of the generator using belts, sprockets, or gears. The large commercially available units generally make a compromise here. They use a relatively low speed generator (1000 rpm) and they gear the generator to the propeller through a small transmission at about a five to one step-up ratio.

The next question is how to decide what size generator to use with what size propeller. Well, here again some compromises are in order. First, you must decide what windspeed will be required for your generator to put out its full electrical output. If you want full output at low wind speeds you will need a large propeller, whereas if you are satisfied with full output only at high wind velocities, a small propeller will suffice.

In general, light winds are more common than strong winds. Statistical studies of wind data show that each month there is a well-defined group of wind velocities which predominate. These are called the *prevalent* winds. There is also a well-defined group which contains the bulk of the energy each month, called, appropriately enough, *energy winds.* The first group, consisting of 5 to 15 mph winds, blows five out of seven days on the average, while the energy winds of 10 to 25 mph blow only two out of seven days. It might seem, at first glance, that you should design for maximum output at, say, 15 mph to take advantage of all those prevalent winds, but this would require a very large propeller for the power produced, and all the power from winds higher than 15 mph would be thrown away.

As an example, consider a 2000-watt generator which is to yield its full output at 15 mph. From Table 1 we can deduce (assuming a 70 percent efficient system) that to get 2000 watts at 15 mph we will need something over a 22-foot diameter propeller. Now this will be large and expensive to build, and difficult to control in high winds. Besides, look at all that power you are throwing away at higher wind speeds. If you really are going to build a 22-foot diameter rotor, then you might as well install a bigger generator on it and get some of that power at higher wind speeds, right?

Well, in practice this is what is done. Most working wind generators are designed to put out full power in wind speeds of about 25 mph, and in so doing they do sacrifice some performance at low wind speeds. Usually they deliver almost no output at wind speeds below 6 or 8 mph, but this is really not a serious drawback because there is so little energy available from these light winds anyway.

Now you can begin to understand just what is meant by a "2000-watt" wind generator. As you can see, it is hard to compare the rated output of a wind generator to that of a conventional generating plant with the same rating. In the case of the wind generator, the power rating merely tells you what the *maximum output* of the generator will be at a certain wind speed—and you must know what this wind speed is if you want to calculate how much power you will actually get from a certain windplant under varying wind conditions.

What Size System?

This brings us to the final problem in choosing a suitable wind electric system for your homestead. The question is, how much total electric energy will a certain size system produce over a period of time in your particular location? This is the main concern of anyone attempting to determine the feasibility of a wind generator in his or her area—and it is also the most difficult question to answer. Suppose you have sat down and, with the help of Tables 3 and 4 you have figured out that, to run everything you want to run in your new wind-powered homestead, you will need 200 kwh (kilowatt-hours) of electricity per month. If all your power is to come from the wind, you will need a system that will provide at least this much per month and a little more besides to allow for the slight inefficiency of the storage batteries.

Well, to actually figure out precisely how much a certain system will deliver in a given location, you must know not only the complete output characteristics of your wind generator at different wind speeds, but also you must have complete windspeed data for the proposed installation site. And by complete, I mean enough data to plot a continuous graph of the wind speed for a year or two. Such a graph would allow you to compute the total energy available from the wind in your particular location. Roughly, the power available would correspond to the area under the curve, but even this is not mathematically correct because of the cubic dependency of power on wind speed. To do it right would require some pretty sophisticated statistical analysis which we will certainly not venture into here—and actually it all becomes pretty academic since few people have good enough wind data for their location, anyway. But, if you do want to know more about this, Putnam's book, *Power from the Wind*[2] will be of some help.

Now, lest you begin to despair, let me give you some idea of how to proceed in the absence of all the facts. This might be considered "fudging it," but unless you're planning a very expensive commercial installation, it will certainly get you into the right ball park. First, find out what the *average* yearly winds are in your location. The Weather Bureau records wind speeds hourly at several hundred stations across the country, and if you write them (see reference 3) you can get this information, including average wind speeds for each month and year at a station near you. Use this as a start, but don't consider it definitive. Winds at your actual site may vary considerably from those at the local weather station, so you will probably want to carry out some tests of your own, especially if you are in a doubtful area, i.e. official average winds much under 10 mph.

Later we will discuss measuring wind speed and selecting the best site in more detail, but for now let's assume you have satisfied yourself that the average winds at your location are, say, 12 mph. Well, we have prepared a handy-dandy little table based on our limited experience in this field, which will hopefully give you some idea of what you can expect from different size windplants at various average wind speeds. As you will appreciate, many factors enter into this and we have had to make several assumptions.

First, these figures are based on typical present production wind generator designs with tip speed ratios on the order of five and efficiencies of about 70 percent. It is also assumed that there is negligible output below wind speeds of 6 mph and that maximum output is reached at 25 mph. This table represents a composite of actual measurements, plus some figures put out by several wind generator manufacturers plus a fair amount of interpolation.

As I said, this table should only be considered as a rough

TABLE 2. Average Monthly Output in Kilowatt-Hours

NOMINAL OUTPUT RATING OF GENERATOR IN WATTS	AVERAGE MONTHLY WIND SPEED IN MPH					
	6	*8*	*10*	*12*	*14*	*16*
50	1.5	3	5	7	9	10
100	3	5	8	11	13	15
250	6	12	18	24	29	32
500	12	24	35	46	55	62
1000	22	45	65	86	104	120
2000	40	80	120	160	200	235
4000	75	150	230	310	390	460
6000	115	230	350	470	590	710
8000	150	300	450	600	750	900
10,000	185	370	550	730	910	1090
12,000	215	430	650	870	1090	1310

estimate of what you can expect from wind–generated power in different wind areas. Many manufacturers of wind generators refuse to commit themselves to anything as specific as the figures listed in this table because they claim that conditions vary so much, what with the effect of turbulence, temperature, etc., that they would only be sticking their necks out to make any specific predictions of long–term energy output. Nevertheless, we feel that this is the one basic statistic that everybody wants to know when they consider installing a wind electric system and so have included this table for your use.

Now we may proceed to use the table to solve the original problem which was, how large a system will you need to get that 200 kwh per month in an area with a 12 mph average wind speed? Checking Table 2 under the 12 mph column we find that a 2000-watt system would produce only 160 kwh while a 4000-watt generator would produce 310 kwh. Interpolating between these two values we can estimate that a 3000-watt unit might produce 230 kwh per month which is just about right, allowing for the inefficiencies of batteries, inverters, etc. Of course, when it comes around to buying or building such a system you may be forced by financial considerations, or by what is actually available, to install a larger or smaller system, but at least you'll have some idea what you can expect from it when it's all done.

Perhaps, as we were talking about maximum outputs at 25 mph and such, you were wondering what happens at higher wind speeds. Well, all modern production windplants are designed to function completely automatically in winds up to at least 80 mph or even higher, so rest assured that there is no such thing as a site with *too much* wind. In order to survive all kinds of winds, wind generators employ some method of holding down their speed in heavy winds. The most common method of spilling excess wind, whenever the power from the wind exceeds the power rating of the generator, is a system of weights mounted on the propeller which act centrifugally to change the pitch of the blades, thus reducing the wind force on the propeller. This system, which amounts to a built-in governor, holds the propeller at a constant speed and prevents overspeeding when there is little or no load on the generator—which happens whenever the batteries are fully charged and no power is needed. This is one area where the modern windplant has come a long way in solving a problem that plagued the windchargers of 40 years ago. Burned–out bulbs and even burned–out generators were not uncommon with the old units as the windmill raced out of control in heavy winds.

But even with the modern version, the manufacturers generally recommend that if wind speeds greater than 80 mph are anticipated, as in a hurricane, the propeller should be manually stopped and/or rotated sideways to the wind. Most models have a brake control located at the bottom of the tower for this purpose. Such "furling" of the windmill during a storm greatly reduces the strain of high wind loads on the propeller and on the entire tower structure.

The Storage System

Power storage is certainly the key to any successful wind–powered electrical system. About the only thing you can say for certain about the wind is that it is always changing; and really, the only reason wind power isn't used more widely is because of its unsteady nature—sometimes it's there and sometimes it isn't. Well, the obvious solution to this problem is to get your power while you can and put it away for when you need it. In theory, at least, there are many ways of doing this. And right now plenty of people are working hard on the problems of energy storage, because they know that this is the key to efficient use of many forms of natural energy which persist in coming in intermittent doses. One of the most promising types of storage for wind-generated electricity, I feel, is conversion of electrical power to hydrogen and oxygen gases by electrolysis of water. These gases may then be stored in tanks and later used to produce power either by direct combustion, or in a fuel cell to produce electricity.

But right now, if we are to proceed to construct an operating wind power system from reliable existing components, we must content ourselves with the good old lead-acid battery. This type of battery has been around for a long time and we know a lot about its behavior. We can predict just how it will act under various conditions and how long it will last. And as of now, it still represents the cheapest practical method of electrical energy storage available for the individual user of wind power.

The storage batteries used in wind systems are similar to ordinary automobile batteries, but they have thicker lead plates and are specifically designed for repeated cycling over a period of many years. This means they can go from a fully charged to a fully discharged state over and over again without damage. The ability to withstand some 2000 complete cycles is typical of this type of battery. They also come with built-in "Pilot Ball" charge indicators which tell you at a glance the state of charge of the batteries. Batteries specially made for this purpose are known as "stationary" or "house lighting" storage batteries and are available in sizes from 10 ah (ampere-hours) to 8000 ah. The smaller sizes (up to 150 ah) usually come as three-cell 6-volt batteries, while the large ones come as single-cell 2-volt batteries. It takes 60 2-volt cells connected in series to produce 120 volts. The number of cells determine the voltage, but the amount of power storage capacity is determined by the size of the batteries and the number of plates in each cell.

The battery set used in each home power installation must be carefully chosen to meet the needs of the individual situation. A typical modern wind power installation will employ battery storage capacity sufficient to meet normal electrical needs for a period of at *least* three days without wind, and often sufficient capacity is provided for as many as seven windless days. Here is a case where you must balance the initial investment of the batteries against the continued smaller expense of fuel for the back-up system. If you opt for only two days of storage you will probably find yourself starting up the gas generator three or four times a month to pull you through the flat spots. But if you can store a week's worth of power, you might not need that back-up system at all.

Power Conversion Devices

And now we get to the technical part. To begin with, there are two basic types of electric generators: the *alternator* which generates alternating current (AC) and the brush type generator which generates direct current (DC). All power companies in the U.S. use alternators which generate 60-cycle per second alternating current. This means that the electricity pulses back and forth, changing its direction of travel in the wire 60 times each second. Since this is the standard power available, all electrical appliances and devices are made to operate on this type of steadily pulsating electricity. So if we are to have a practical power source for everyday household use, we must somehow supply this type of alternating current, and at the standard voltage, which is about 115 volts.

Now this poses a definite problem for a wind–generating system because 60-cycle alternating current is produced only by an alternator turning at a constant speed, usually 60 revolutions per second. But since the wind never blows at a constant speed, you cannot get a constant number of revolutions per second from a wind–driven alternator. Instead, you get an rpm which varies with the wind speed. (The main reason power companies provide us with AC is for *their* convenience; AC can be transmitted along power lines more easily.)

One solution to this problem is to go ahead and use an alternator to generate the power (because it is more efficient and lasts longer than a generator) but don't worry about the uneven rpm. Pass this

AC through rectifier diodes and convert it to a steady DC current which can then be sent directly to the batteries for storage. Now, to get that steady 60-cycle AC we've been working up to, you can use another alternator run by a DC electric motor which can run off all that nice DC current you just stored in your batteries. This DC will not fluctuate and so your alternator will turn at a constant rpm! Such a device is called a "motor-generator" or a "rotary inverter," and it produces electricity which is indistinguishable from the Power Company variety.

The reason I mention this is that there are other ways of converting DC to AC but often these produce a "square wave" instead of a sine wave, which is what the Power Company makes. You need the sine wave for static-free stereos and distortion-free television sets. In other words, many electronic devices are sensitive about the shape of the wave.

A "wave" is produced when you make a graph of the voltage on one axis and of the time on the other axis. It shows you graphically how the voltage (or current) is fluctuating with passing time. A graph of DC current would be just a straight line. (I told you this was going to be the technical part!)

The mechanical rotary inverter is only one way of converting DC to AC. Although it does a good job, it is not very efficient—on the order of 60 percent. There are now available many types of electronic, solid–state inverters, which perform the same function as the rotary inverter, but more efficiently—about 80 percent is typical for these units. They contain no moving parts at all and many of them produce an approximation of a sine wave which is good enough for all practical purposes. The only problem is that they still tend to be more expensive than the old fashioned rotary inverters which are available as army-navy surplus items.

One final remark about power conversion. As it turns out, if you have chosen a convenient voltage for your whole system (such as 12 or 115 volts), you can use much of your power directly from the windmill and batteries as DC, and not put all your power through an inverter. Since the inverter uses some power just to operate, you want to use it as little as possible and save that power. Light bulbs, all devices with simple heating elements, hand power tools, and other appliances with universal AC-DC motors, will work very nicely right off your DC with no conversion, as long as the voltage is right.

(This is one important reason for choosing a 115-volt system.) Now you can save your inverter for the record player, electric typewriter, radio, or television. This is what we do.

Our Quirk's Installation

Speaking of what we do . . . We presently have in operation at our homestead in North Orland, Maine, a complete wind electric

These 2 wind generators, an Australian Dunlite and a Swiss Elektro, supply the Clews homestead with the power to run most appliances.

system which provides *all* of our power for lights, power tools, appliances, and water pump. The heart of this system is a 2000-watt, 115-volt, Quirk's "Brushless Windplant" mounted atop a 40-foot steel tower out behind our house. These components, as well as the 130-ah houselighting batteries which provide our storage for windless periods, are all standard production items which we obtained from Quirk's in Australia and which we now can offer for sale to others through the Solar Wind Company.

Our Quirk's Windplant consists of a large 115-volt, three-phase AC, low speed alternator which is attached through a gearbox to a propeller hub which holds three long slender aerodynamically shaped propeller blades. The diameter of the propeller is 12 feet. The whole unit—generator, gearbox, and propeller—is mounted on a free-swivelling platform onto which a tail has been attached to keep the propeller blades facing into the wind. Also contained in the swivelling base are three silver-coated slip–ring commutators which conduct the electricity from the generator above, down to the fixed tower top below.

The Quirk's Company sells windplants in several different models. The one we chose is their largest standard production model and is one of two alternator type models offered. The prime advantage of an alternator is that it has no brushes to wear out and to replace. The *only* maintenance required on these units is an oil change of the gearbox oil (one quart) once every five years. Quirk's claims that several of these windplants have been in continuous operation for over 40 years without need of service. It is hard to imagine any engine–driven generating plant giving such service. In fact, most common models require a complete overhaul over 1000 hours of use—which figures out to only 42 days of continuous use! This is something to consider when comparing costs of wind electric systems to conventional systems—not to mention the costs of fuel for the engine–driven rig.

Our windplant begins to turn in about 5-mph winds and actually starts charging the batteries in 9-mph winds. Outputs at various wind speeds, from our measurements, are: 10 mph, 60 watts; 15 mph, 600 watts, 20 mph, 1200 watts; and 25 mph, 2000 watts or the full rated output of the generator. This data shows close agreement with the theoretical output of a windmill with 70 percent efficiency (see Table 1).

At full output the propeller is only turning at 150 rpm, while the generator which is geared to it at a 5 to 1 ratio, turns at 750 rpm —or about the idling speed of an automobile engine. The slow speed of the Quirk's generator is responsible for its long life. For the propeller, the slow speed is not only desirable, but necessary. It is desirable because the efficiency actually decreases when tip speeds begin to exceed 100 mph, and it is necessary because large, and potentially destructive, centrifugal loads can build up in a large propeller at high speeds. The huge 175-foot diameter wind generator on Grandpa's Knob in Vermont developed its full power at a propeller speed of only 28 rpm!

Storing Our Power

In our Quirk's generator, the AC current which is generated is converted to DC by diodes located right inside the alternator housing, so that it is DC current that flows down the wire to our house 250 feet away from the tower. This current goes directly to a transistorized control panel which comes as part of the Quirk's system. This panel, which includes a large voltmeter and ammeter to show you exactly what is going on at any given time, automatically channels the current to where it is needed. Current from the generator can flow right through the panel and directly to any device which is switched on at the time. If there is more than enough current, any extra is automatically channelled to the batteries for storage. If the batteries are fully charged and no power is needed, the control automatically cuts off the generator output (by lowering the voltage to the field, for those of you who really want to know).

So the net effect is that you operate directly from the windmill whenever there is sufficient wind, but when there is not, the batteries take over and supply the extra power, or all the power when there is no wind. But it is all automatic; the only way to tell whether you're running on wind or batteries at a given moment is to check the ammeter which registers current flowing in or out of the batteries. The batteries serve as a stabilizer for the fluctuating wind-generated current. There is no noticeable variation of light intensity with varying wind strength, and even when large appliances are switched on, the lights remain steady. With DC there is no flickering that you sometimes notice with AC lights, and light bulbs last much longer too.

An easy way to visualize this whole system is to think of a water analogy. Consider that the windmill is a water pump which, depending on the wind, can pump from zero to a maximum of 100 gallons per hour of water. The batteries are represented by a big water tank capable of holding, say, 1000 gallons of water. If we begin with an empty tank (dead batteries) and do not draw any water (current) it will take ten hours for the pump running at full capacity to fill the tank (about the same time it actually takes to charge our batteries from a fully discharged state). Now let's turn on some water, say, 50 gallons per hour. This will come directly from the windmill, with the tank just standing by. This is a very common situation with our system. Suppose, however, the wind is blowing such that we're only getting 25 gallons per hour into the tank. If we draw off water at the rate of 50 gallons/hour, 25 will come from the windmill while an additional 25 will come from the tank. This also is quite common, especially during the evening hours when winds are often light. But since the windmill is left on 24 hours a day it can make use of wind at any time to replace water (electricity) used at an earlier less windy time.

Now you begin to see the importance of the batteries. The more and the larger the batteries, the bigger the tank and the better you can maintain continuous power through periods of fluctuating winds.

Our battery set consists of 20, 6-volt glass-encased lead acid batteries connected in series, giving a total output of 120 volts and having a storage capacity of 130 ah. This means you can draw 1 amp for 130 hours or 10 amps for 13 hours, or anything in between. In practice this seems to give us about four days of power without wind. In all the time that the system has been in operation, we have only had to run the auxiliary gasoline generator a total of 20 hours to maintain continuous power. This has consumed 5½ gallons of gasoline at a total cost of $1.75. Quirk's recommends that if you are really dependent on power, for instance running a refrigerator or freezer, that you should have a back-up power source such as a small gasoline generator. Ours, purchased from Sears, is a 1600-watt model that cost about $200. This 115-volt AC alternator can be used to provide auxiliary power as well as to charge the batteries by connecting it through a bridge rectifier which converts AC to DC. The rectifier we use is a Motorola MDA 990–3 which can be obtained at any electronics supply house for a few dollars.

Our batteries are "Century House Lighting" batteries made by the Century Storage Battery Co., Ltd. in Australia and distributed through Quirk's. They come in completely transparent polystyrene cases and have built-in "gravity ball" indicators which tell you at a glance their state of charge. In our house, they are mounted in a long row under a bench which can be folded up to inspect them. So far they have required no attention whatever. Quirk's claims that the average life of these batteries used in this application is 15 to 20 years.

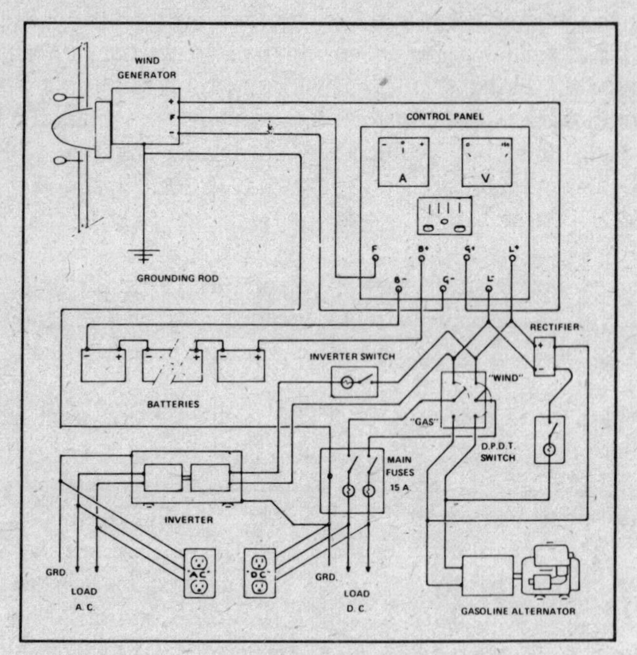

FIGURE 1. Electrical schematic of the Clews homestead.

Our Power Conversion Devices

We initially obtained a 250-watt rotary inverter as a Navy surplus item from a company near Boston.[4] This is a motor-generator type unit like the one described earlier, and it has built-in controls which maintain the voltage and 60-cycle frequency constantly no matter what load is being drawn. This inverter gives us perfect TV, radio, and stereo operation, and it provides a good source of steady 115-volt AC "house current" for all small appliances which use 250 watts (2.2 amps) or less.

What we have done is to run two sets of wires in our house,

and in most places, put two outlets side by side, one labeled "AC," the other "DC." The DC outlets are always live, but the AC outlets come to life only when the inverter is turned on—which is now done by flipping a switch near the control panel. Someday I will set this up so that the inverter will come on automatically whenever an AC appliance is turned on—this could easily be done with a relay.

As it is now, since the inverter draws some current just to run it, it is switched on only when needed. But even doing this, we found that it was quite wasteful to use our original 250-watt inverter for such items as the electric typewriter and solid-state stereo. These draw less than 50 watts apiece, but were often on for long periods of time. When run through the inverter, these appliances were drawing about 150 watts DC or about three times their normal AC consumption. The reason for this is that mechanical inverters are very inefficient at small loads. In fact, even when there is *no* AC load on the inverter, it draws 100 watts of DC power from the batteries just to run itself—this is called "no load" power consumption.

Well, the best solution to this is an electronic solid state inverter. These are available in all sizes up to 10,000 watts, and they have very low, no-load power consumptions. Even a 3000-watt model, which would be adequate for almost any home appliance requiring AC (such as refrigerators, freezers, workshop equipment with up to ¾ hp motors, etc.), draws only 60 watts at no load. But these are expensive, and seem to represent something of an "overkill" for our typewriter problem. So, we bought another smaller rotary inverter rated at 75 watts maximum output which uses only 15 watts at no load. Using this inverter, the typewriter and hi-fi draw only 50 watts of DC power. So you can see that by matching the size of the inverter to the load, you can drastically cut down losses, even using the relatively inefficient, but cheap, mechanical rotary inverter. The electrical schematic below will show you how our system is set up at present.

We have really just begun to discover the possibilities in this business. It would seem that if you have the money and the inclination, you can do just about anything you want with a wind-powered, battery-operated electrical system. For example, there is actually available an inverter system that will pick up the "vibrations" of the Power Company's 60-cycle current up to 10 miles away and automatically keep *your* power tuned to this 60-cycle frequency. You could even run an electric clock on something like this! There is also plenty of equipment available that will handle automatic switching of your power system from the power lines to a wind electric system, for those who might want to install such a system as an emergency back-up in case of power failure.

One of the most interesting possibilities, which we have heard has been tried experimentally, is to use wind-generated power in a location where there is access to the Power Company lines, and to actually put power back into the lines during windy spells thus running the electric meter backwards and reducing the electric bill! This would be something like having 100 percent efficient, infinite capacity storage batteries. But while there is always room for improvement, our system at present works well. We have literally had no problems since the windmill was turned loose last August, 1972 —it just seems to keep going around, making free electricity. I guess this is what you expect when you lay out over $2500 for a production machine.

Installation

As for installation costs, we did this all ourselves, so costs were negligible. We used cardboard "Sonotube" forms and filled them with concrete to form three, 8-foot pylons buried 3 feet in the ground. There is 1 cubic yard of concrete in each pylon. The tower comes in all these little pieces and must be bolted together—the biggest Erector Set you ever saw! We chose to set the base first and build the tower up from the ground, which is not too hard if you're not scared of heights. The top is as high as a five-story building. (Retta, my wife, adds here that the scary part was for the the helpers on the ground who were constantly dodging falling tools!)

The only problem with building the tower from the ground up, as we discovered to our chagrin, is the setting of the windplant—a mere 400 pounds—on top of the completed 48-foot tower. Of course you could hire a crane, but that's no fun. What we did was much more interesting. After several false starts, again involving falling objects and bad tempers, we got a wooden tripod on top of the tower, with a pulley under the tripod's peak. We passed a stout rope through the pulley and attached one end to the 400-pound windplant and the other end to a 4000-pound Land Rover. Retta drove the Land Rover slowly away from the tower while the kids sat under a nearby tractor for protection and shouted directions. Up went the windplant—oh yes, with a guy wire attached to a brother-in-law—and I, waiting at the top to receive it, wasted no time in bolting it quickly in place. Mostly it worked out OK and it proved that the whole installation *can* be done without expensive equipment or hired professional help.

In the second installation of an identical unit, some 35 miles south of here, they did it differently. There, they assembled the entire tower lying down. The complete windplant was attached, and even wired, and then the whole rig was pulled up into position by a small crane hired for the job. Although this cost a little more than our method, the entire job was completed, and the windmill was put into operation, within 10 days after the equipment was unloaded from the freighter in Boston!

This installation, at the home of Mr. Neil Welliver of Lincolnville, Maine, is the second Quirk's system to be put into operation in the United States. It is being used to replace a 5000-watt L.P. gas–generating plant which previously supplied all their power at a cost over $50.00 per month (plus many headaches with service and maintenance problems). A wind electric system may have trouble competing financially with Power Company power, but it really comes into its own when compared with any other type of independent power plant. Mr. Welliver figures that, compared to his gas generator, the wind generator installation will have completely paid for itself in three to four years time. From then on the power is FREE!

The Pros and Cons of Our Power System

But there are always skeptics. We have received many letters saying, "Sounds neat, but how does it work out in practice?", or, "If it is such a great idea, why aren't other people doing it?" Well, the fact is that quite a few other people *are* putting wind power to practical use. In the mountains of Switzerland there are several hotels and restaurants with modern electrical equipment which rely completely on the wind for their electrical power. And near Albuquerque, New Mexico, Robert Reines lives in a house which is completely powered by the wind—and heated by the sun! And there are many others. We have received many letters describing wind electric systems either planned, or already in operation, all through the U.S. and Canada.

And then, there's us. We are a family of four living quite comfortably in the wilds of Maine with only the power we get from a single Quirk's windplant. How does it work out in practice? Well, so far it's done all we've asked it to do. That is, provide power enough for six or eight 75-watt light bulbs every evening, radio, T.V., stereo, electric typewriter, blender, toaster, vacuum cleaner, skill saw, electric drill, as well as for a ⅓ hp deep-well water pump. Compared to life before we had power at all, it's heaven. An old hand pump may look romantic, but it ain't!

In our main fuse box we have 15-amp fuses, so I guess if you want to compare our wind electric system to the Power Company service, you could say we have a 15-amp service. But since the tools are used during the day and the lights at night, we've had no problem with overloads as yet. We estimate that with our system we are getting about 110 kilowatt-hours of electrical energy per month. This is on the order of one–fourth the amount of power that the average American family consumes from the Power Company. If you check Tables 3 and 4 you can get a better idea of what 110 kwh of power really means in terms of various home appliances. A refrigerator, for example, normally requires about 95 kwh per month, and as you can see, our system could not supply this much more in addition to our present load. So, even if we had an inverter large enough to handle a refrigerator, right now we do not have the extra power to run one. At present we use an L.P. gas refrigerator which works very well, but we are already planning to add additional wind-powered generating equipment to provide for an electric refrigerator and freezer. But in general we are very pleased by our present system and continue to be amazed at its day-in-day-out trouble-free operation through rain, sleet, and snow. Even ice storms and winds up to 60 mph have had no adverse effect on our wind plant.

A wind electric system does have some limitations—that is, unless you are willing to go all out for a really large system. Most systems do not provide 230 volts for large appliances such as electric ranges, clothes dryers, air conditioners, baseboard heat, etc. It is possible to obtain 230-volt wind generators on special order, but generally the items which require this voltage also use large amounts of power and require a large and expensive system.

One approach to this problem is to use *two* large 115-volt wind plants to provide both 115-volt and 230-volt power using a dual set of batteries as well. A complete system of this type using dual 6000-watt generators, 500-ah batteries, a 3000-watt dual voltage inverter, etc., might cost about $9000 and would provide up to 1000 kwh of power per month in a good wind area. Now something like this *would* provide enough power for an electric range, electric hot water heater, electric dryer, and even some electric heat. And, just think, it's all free after the initial installation!

But our present system isn't quite so elaborate, and for the time being anyway, we are content to use wood for heat and cooking in the winter and bottled gas in the summer. And for a clothes dryer we use a special "solar-wind" model, which you too can set up in your own back yard for less than one dollar! (Think about it.)

Wind Speeds and Site Selection

Our location is about 15 miles inland from the Atlantic coast, roughly on a line between Southwest Harbor and Bangor, Maine. Our elevation here is 350 feet above sea level, but we are not located on the highest ground around by any means. There is a 950-foot mountain a half mile to the southwest and an 850-foot hill a quarter mile to the northeast. But in spite of these wind blocks we still get winds which seem to average 10 mph, which is exactly the published average wind speed at our nearest weather station, the Bangor International Airport. Of course, it's not the average winds that give you power, but the 15 to 20 mph breezes that really count. I would guess that we get two to three of these per week for a period of at least six hours.

There is no doubt that the immediate topography has a very

important effect on the winds. You should not rely too heavily on statistics for nearby areas, but instead carry out some tests of your own. And in actually selecting the exact site for the windplant, you might want to make some careful studies of wind behavior in your particular area. For example, we have found that while maximum height is generally desirable, locations even fairly high up on the side of a hill (especially the east side) may not be as good as a location much lower but farther away from the hill. Usually if you cannot get right on top of a hill, it is better to stay away from it altogether.

If you want to find out more about the winds in your location you might try the following. Get yourself a wind gauge,[5] and go out and take a reading every day at the same time, preferably in the afternoon, in the location or locations that you consider best for your windplant. Of course this may mean climbing a 50-foot tree every afternoon at 3 P.M., but the exercise will do you good! If getting up into the air is a problem, try to find a location nearby that will allow you to get up into clear air. I would think that if the results of your tests show that there is over a 10-mph wind an average of two to three days per week you have an adequate site for wind power.

You might also wish to compare your own local readings with those of your closest weather station. If, after you have faithfully recorded wind speeds every day for a month, you go visit your local weather station, you can compare your figures to theirs taken at the same times. In this way you should be able to form a direct correlation between your figures and theirs and thus establish a relationship that will allow you to apply the long-term wind data collected by the Weather Bureau over many years to your particular location.

Remember that even in an open field with no trees around, the wind speed at 30 feet above the ground is always 20 to 50 percent stronger than at the surface. I think that any successful large wind generator should be mounted at least 30 feet above the ground and additionally about 10 feet above any surrounding objects, trees, buildings, etc., within a 500-foot radius. Distances up to 1000 feet away from your house are acceptable, but the closer, the better, and the cheaper the lead-in wire will be.

Building Your Own Wind Generator

So far we have discussed wind generators mainly in terms of those which are commercially available. And if what you want is a complete, reliable installation which will produce relatively large amounts of power for many years to come without problems, you should probably consider one of these production units. There are wind generators currently in production in various parts of the world ranging in output from 50 watts to 25 kw (25,000 watts).

But there are many people who are interested in experimenting with a home-built design, and in bringing down the cost of the components by using recycled components and building many parts themselves. At present, there is no easy way to build your own windplant. There are no complete kits or plans available, to our knowledge. But several people are working on this right now, ourselves included, and I would venture to guess that in the near future such information will be available.

Most home-built designs rely on the common automobile alternator for power generation. These are certainly the cheapest generators available, considering their output, and are readily available. But there are two basic problems which must be overcome with the automobile alternator. First, it is designed to deliver its power at the relatively high speed of 2000 to 5000 rpm. So it must be geared to the propeller by belts, gears, or chains and sprockets. Gear ratios of at least 5 to 1 and as high as 10 to 1 are used in most successful designs.

The second problem is that, in order for the automobile alternator to start charging, the field must be energized; (in your car this happens as soon as you turn on the key). But in a wind generator this means that someone has to be around to turn it on when the wind is blowing, and off when it is not, or it will quietly sit there and drain your batteries during calm spells. One way to overcome this problem is to incorporate a small wind-sensitive switch in your unit —just a microswitch with a little wind-paddle attached to it, that automatically turns it on at wind speeds over 7 mph, or at whatever speed your windplant begins charging.

Most of the home-built designs using car alternators also use auto voltage regulators as well and operate as 12-volt systems with

12-volt batteries. This is actually quite practical as there are many devices available which are built to operate on 12 volts DC, including light bulbs, radio and stereo equipment, water pumps, and even inverters that will step the voltage up to 115 volts AC. Electronic inverters of this type are mass produced (for campers, etc.) and are relatively inexpensive. But something many people are not aware of is that the common auto alternator will also produce 115 volts. All you have to do is throw away the regulator and disconnect one output terminal from the battery (but leave the field wire connected to a 12-volt source). Now if you spin the alternator up to 4000 rpm you can get 115 volts and a vastly increased power output. Of course the alternator diodes may burn out on you, but they can be replaced with larger ones quite inexpensively.

Many people are presently working on their own designs and some of them are coming up with some very ingenious solutions to the problems of a practical and cheap wind generator. Perhaps you can make a contribution to this field. But, before you begin, arm yourself with as much knowledge as you possibly can about this complicated subject. And remember that your completed design must be capable of withstanding the strongest winds you will ever get in your area. Don't be one of the people who writes us that they had a windmill that worked "beautifully" for a few days only to be demolished by the first strong winds that came along. Apparently, this is not an uncommon experience. The information that follows should help you get started.

Calculating Your Power Needs

Here is some data to help you figure out how much power you will really need in your wind-powered homestead. First, a basic electrical formula that will help you juggle watts, volts and amps around.

$$\boxed{\text{watts} = \text{amps} \times \text{volts}}$$

From which follows: amps = watts / volts

And: volts = watts / amps

Watts are a measure of power. A 75-watt bulb requires 75 watts of power to run it. If it is a standard 115-volt bulb, then it will draw 75 / 115 or .65 amps of current. If it is a 12-volt, 75-watt bulb, however, it will draw 75 / 12 or 6.25 amps. This illustrates that more current is needed to get the same power at a lower voltage. This is important to keep in mind because the size wire required in a given application is determined solely by the *current* which will be flowing through it, and not the power.

So, if you're running a wire which will power four 75-watt bulbs, for example, at 115 volts this would only represent 2.6 amps, whereas at 12 volts it would be 25 amps and might require special heavy-duty wiring. This is also something you'll want to consider when choosing a voltage for your system.

Another relationship you will want to understand is the connection between kilowatt-hours and ampere-hours (ah) and how this measure of total energy consumed relates to the volts, amps, and watts of a specific appliance in your house. Batteries are usually rated in ah capacity. This tells you how many amps they will deliver for how long. A 100-ah battery will deliver 1 amp for 100 hours, or 10 amps for 10 hours, etc. An ah is just what it sounds like: amps times hours. But as we have seen, the current in amps that a certain item draws depends upon the voltage at which it is operating. So an ampere-hour rating in itself doesn't tell you very much about how much power you can store in the batteries.

In order to determine this, you must also specify a voltage—such as 100 ah at 115 volts. Now, since we know that amps times volts equals watts, we can infer that ah times volts equals watt-hours, which in fact it does! And 1000 watt-hours equals one kilowatt-hour, and this is what the Electric Company bills you for every month. A kilowatt-hour is nothing more than 1000 watts of power used for a period of one hour (or 500 watts for two hours, etc.). Now if this isn't all crystal clear, a careful study of Tables 3 and 4 should help.

Here is how you can use Table 3 to figure out what size system you'll need. First, prepare yourself a similar table, listing only those items you will use in your wind-powered home. The table will be most accurate if you substitute actual figures for the average values listed in the table. For example, if you can read directly off the label on your toaster that it draws 1000 watts instead of the 1146 listed

TABLE 3. Power, Current, and Monthly KWH Consumption of Various Home Appliances

APPLIANCES	POWER IN WATTS	CURRENT REQUIRED IN AMPS at 12v	CURRENT REQUIRED IN AMPS at 115v	TIME USED PER MO. IN HRS.	TOTAL KWH PER MO.
Air Conditioner (window)	1,566	130	13.7	74	116
Blanket, electric	177	14.5	1.5	73	13.
Blender	350	29.2	3.0	1.5	0.5
Broiler	1,436	120.	12.5	6	8.5
Clothes Dryer (electric)	4,856		42.0	18	86.
Clothes Dryer (gas)	325	27.	2.8	18	6.0
Coffee Pot	894	75.	7.8	10	9.
Dishwasher	1,200	100.	10.4	25	30.
Drill (¼ in. elec.)	250	20.8	2.2	2	.5
Fan (attic)	370	30.8	3.2	65	24.
Freezer (15 cu. ft.)	340	28.4	3.0	290	100.
Freezer (15 cu. ft.) frostless	440	36.6	3.8	330	145
Frying Pan	1,196	99.6	10.4	12	15.
Garbage Disposal	445	36.	3.9	6	3.
Heat, electric baseboard, ave. size home	10,000		87.	160	1600.
Iron	1,088	90.5	9.5	11	12.
Light Bulb, 75-watt	75	6.25	.65	120	9.
Light Bulb, 40-watt	40	3.3	.35	120	4.8
Light Bulb, 25-watt	25	2.1	.22	120	3.
Oil Burner, ⅛ hp	250	20.8	2.2	64	16.
Range	12,200		106.0	8	98.
Record Player (tube)	150	12.5	1.3	50	7.5
Record Player (solid st.)	60	5.0	.52	50	3.
Refrigerator-Freezer (14 cu. ft.)	326	27.2	2.8	290	95.
Refrigerator-Freezer (14 cu. ft.) frostless	615	51.3	5.35	250	152.
Skill Saw	1,000	83.5	8.7	6	6.
Sun Lamp	279	23.2	2.4	5.4	1.5
Television (B&W)	237	19.8	2.1	110	25.
Television (color)	332	27.6	2.9	125	42.
Toaster	1,146	95.5	10.0	2.6	3.
Typewriter	30	2.5	.26	15	.45
Vacuum Cleaner	630	52.5	5.5	6.4	4.
Washing Machine (auto)	512	42.5	4.5	17.6	9.
Washing Machine (ringer)	275	23.	2.4	15	4.
Water Heater	4,474		39.	89	400.
Water Pump	460	38.3	4.0	44.	20.

in the table, then use this figure in your table. To find the current for your toaster at 115 volts use the formula, a = w / v to get 1000 / 115 or 8.7 amps. You may also wish to revise the average times that the appliances will be used each month according to your own use patterns. The values listed in the table are statistical averages for a family of four living with unlimited power from the Power Company.

Now, to find the actual kilowatt-hours per month, multiply watts times hours used per month (this gives you watt-hours per month) and then divide by 1000, by moving the decimal point three places to the left, to get kilowatt-hours per month.

From Table 4 we can tell several things. First, we can see that to supply all our electrical requirements we will need about 95 kwh

TABLE 4. Sample Power Consumption Table For Clews Homestead

APPLIANCE	POWER IN WATTS	AMPS @ 115V	TIME USED PER MO. IN HRS.	TOTAL KWH PER MO.
Blender	350	3.0	0.5	0.2
Drill—¼ in.	250	2.2	0.5	0.1
Lights—eight 75-watt bulbs	600	5.2	120.0	72.0
Sabre Saw	325	2.8	0.5	0.1
Stereo	50	0.4	50.0	2.5
Skill Saw	1000	8.7	6.0	6.0
Typewriter	40	0.35	15.0	0.6
Vacuum Cleaner	630	5.5	1.2	0.8
Water Pump—⅓ hp Jet Pump	420	3.6	30.0	12.5
	Total Circled amps	12.3	*Total kwh*	94.8

of electrical energy per month. Going back to Table 2 we find that a 2000-watt system operating in an area with 10 mph average winds will produce about 120 kwh per month, so we're OK (which we know anyway because we've been doing it since August 1972), but this shows that our current usage is fairly near the natural limits of our system in this particular area.

Secondly, from Table 4 we can also figure out what size batteries we will need. There are two things which determine what size batteries are needed in a given application. These are (1) total energy storage required, and (2) maximum current to be drawn at any one time. The first condition is usually the more important. In our case (see Table 4) our total monthly consumption is 95 kwh per month. Dividing this by 30, we can find that the average *daily* consumption is about 3.2 kwh which is the same as 3200 watt-hours per day. (One kilowatt equals 1000 watts.) Now to convert this to amp-hours, we use the formula a = w / v, or ah = 3200 watt-hours / 115 volts or 28 ah. This means that our daily consumption is equivalent to 28 ah at 115 volts, and if we want to provide storage enough for at least four days, we need four times 28, or at least 100 ah worth of batteries. In our case we actually have batteries rated at 130 ah, so again we're OK.

The second consideration in choosing a battery set is the maximum current that will be drawn. House lighting storage batteries are designed to give relatively low currents for long periods. Our 130-ah set, for example, is rated at 15 amps maximum discharge rate. (Automobile batteries often deliver over 100 amps during starting). This low current rating is part of the reason that house lighting batteries have such a long life. But this means that during periods with no wind, you must limit yourself to the current rating of your batteries—in our case 15 amps. This is why we have 15 amp fuses in our main fuse box. In Table 4 we have circled the items which might be operating simultaneously to produce the maximum load, and as you can see this adds up to 12.3 amps, which again is OK.

Finally, here is a table (Table 5) which lists the average power requirements for various sized electric motors. DC motors are available in all sizes and voltages (see Appendix III) and represent the most efficient means of obtaining power from a wind electric system. Many common appliances can be converted to DC simply by replacing the AC motor with a DC motor which matches the voltage of your installation. In this case you need only concern yourself with

TABLE 5. Power Requirements of Motors in Watts

ELECTRIC MOTOR SIZE	RUNNING	STARTING		
		Induction	*Capacitor*	*Split Phase*
1/6 HP	275	600	850	2050
1/4 HP	400	850	1050	2400
1/3 HP	450	975	1350	2700
1/2 HP	600	1300	1800	3600
3/4 HP	850	1900	2600	—
1 HP	1100	2500	3300	—

the "running" loads, as the batteries are easily capable of absorbing the transient starting loads. If, however, you are considering running AC equipment through an inverter, and this may be the only practical way to run certain items such as modern refrigerators, etc., then you must consider the starting loads as well, because most inverters are not capable of handling even temporary overloads.

So there you have it. As you can begin to appreciate, the business of designing a good wind electric system is no simple matter; but I think you will find it to be a richly rewarding endeavor. We have found that a modern, well designed, wind-generating system can be a source of constant delight.

Do-It-Yourself Wind Generators

James B. DeKorne

Whenever anyone tells me that he or she wants to install a wind generator on a homestead, my first advice is: *Buy a new one if you can afford it.* Unfortunately, most of us who have returned to the land are in no position to lay out several thousand dollars for a new wind generator.

My second best advice is: *If you can't afford a new machine, try to locate a used one and rebuild it.* Just about anywhere in the rural Great Plains states wind generators were common sights on farms during the 1930's and 40's. When the Rural Electrification Administration finally strung its wires to these localities, the generators were often taken down and sold for scrap. Occasionally, however, after a lot of back road driving and many conversations with farmers old enough to remember them, one can find a few of these generators still standing. (Usually they weren't taken down because most folks aren't into the hair-raising job of working with upwards of 500 pounds of machinery while tied onto the top of a high tower—an experience roughly analogous to removing an engine from an automobile while 45 feet up in the air!)

The two most common wind generators were the Wincharger and the Jacobs—respectively the Chevrolet and Rolls Royce of homestead wind-electric plants. The Wincharger came in several models, from 6 to 110 volts, and from 200 to 1500 watts. The Jacobs, a much heavier machine, was built in 32- and 110-volt configurations, and ranged from 1500 to 3000 watts. The 32-volt models of both makes were the most popular in their day, and of the two makes, the Wincharger is the brand you're most likely to encounter now. The Jacobs is quite rare nowadays, but a real find if you turn one up in reasonably good condition.

Almost any old wind generator you may locate is more likely than not in need of extensive restoration. It is unusual to find one that still has usable rotors (blades)—these, being made of wood (with the exception of some later models of the Wincharger, which had aluminum rotors) are the first parts to deteriorate. After all, the machine has probably stood untended for well over 25 years of summer thunderstorms and winter blizzards! It is even rarer yet to find one that still has the original control box, though these can be made up by most any electrician worth his salt.

Is it worthwhile to try to restore one of these old wind-electric plants? Most definitely yes! If you are selective, and can locate a machine that doesn't have irreparable damage, such as broken castings or major parts missing, a little bit of enjoyable restoration will reward you with up to 3000 watts of "free" electricity. Anyone who has rebuilt a Model A Ford, or likes to fool around with old cars, will feel right at home tearing into a 1940's vintage wind generator.

Unfortunately, however, used wind generators in rebuildable condition aren't all that easy to find. That leaves us with choice number three—home-built wind generators. It is my opinion that most of the do-it-yourself machines that I have seen and read about weren't worth the time and money spent to build them. Does this mean that homemade machines aren't practical? Not necessarily so —in the following pages I will attempt to describe the limitations of these machines and outline possible ways in which these limitations might be overcome.

How Big a Wind Generator?

It has been estimated that the average American home uses 10 kilowatt hours of electricity per day.[6] That's about 300 kwh per month. The Clews homestead uses 110 kwh per month with a new 2 kw, 115-volt Quirk's wind generator. Most home-built wind generators are made from 12-volt automobile alternators or generators which are rated at about 500 watts or ½ kw. If a new 2-kw machine gives roughly 100 kwh per month, then it follows (assuming similar wind conditions) that a ½ kw machine will give about 25 kwh per month, or less than 1 kwh per day and less than one-tenth of the American average consumption. And there you have limitation number one—home-built wind generators *using automotive components* (assuming they can be constructed to operate at peak efficiency—which we shall see is a big assumption), still don't put out enough power to equal one-tenth of what most Americans consider "normal."

Now, I would be the first person to say that we should all revise our conception of what is "normal" electrical consumption—lest anyone should take me to task, let it be known that I have deliberately lived without any outside source of electricity at all for over three years. The point I am trying to make here is that if we are going to go to wind-generated electricity, we should shoot for something capable of powering more than a few 25-watt bulbs. We are, after all, advocating an enlightened technology, not a return to the 19th century.

Which Generator?

I have always felt that (with the exception of specialized applications) wind generators of under 1000-watt capacity are impractical for most people's needs. Since electrical devices designed for automotive usage are not really adequate for the needs of a modern homestead, we can look more profitably in the direction of surplus aircraft equipment to build our wind generator. For example, on page 57 of the Palley Supply Co. catalog (see Appendix III), a firm dealing in surplus military and industrial equipment of all sorts, we see listed (among many others) a surplus aircraft generator (#G–1273–1A) which produces 75 amps at 24 to 32 volts. Since volts × amps = watts, we deduce that this unit is capable of putting out about 2400 watts. (That's 400 more watts than a standard Quirk's machine!) This unit weighs only 31 pounds.

In using the Quirk's unit as a comparison, I don't mean to imply that this aircraft generator is "better" because it puts out more power—it almost certainly is not of comparable quality since, for one thing, it weighs but a fraction of the Quirk's machine. Thirty-one pounds compared with 400 pounds indicates that the Quirk's Company engineers designed their unit for a lifetime of efficient, trouble-free operation *as a wind-driven machine,* and that the aircraft generator was designed for high output at high rpm and minimum weight. Since aircraft are torn down periodically for maintenance and replacement of parts, the above generator was not expected to last for more than a specified number of hours of operation.

This is limitation number two of a homemade wind electric system—one is of necessity required to make use of generators which were never designed for the unique conditions of wind power. The same conditions hold true for automotive adaptations, of course, so we're still ahead by going to the higher wattage aircraft generator; we just can't expect it to hold up as long or as well as a unit designed specifically for wind use.

The aircraft generator puts out its rated power between 1800 and 2500 rpm, so we know that we are going to have to gear it up. This is necessary because even a well-designed and carefully built rotor ("propeller") will turn at only about 150 to 300 rpm. This confronts us with limitation number three: "the maximum power that a wind turbine can get from the wind passing through the disk area of its blades is 57 percent of the wind energy."[7]

It has also been estimated that any gearing device will eat up another 15 percent of the available power. These limitations, of course, are true for *any* wind generator, but it behooves us to be aware of them so that we can design our unit to make efficient use of the 42 percent of (free) power remaining to us. Clearly, the rotor

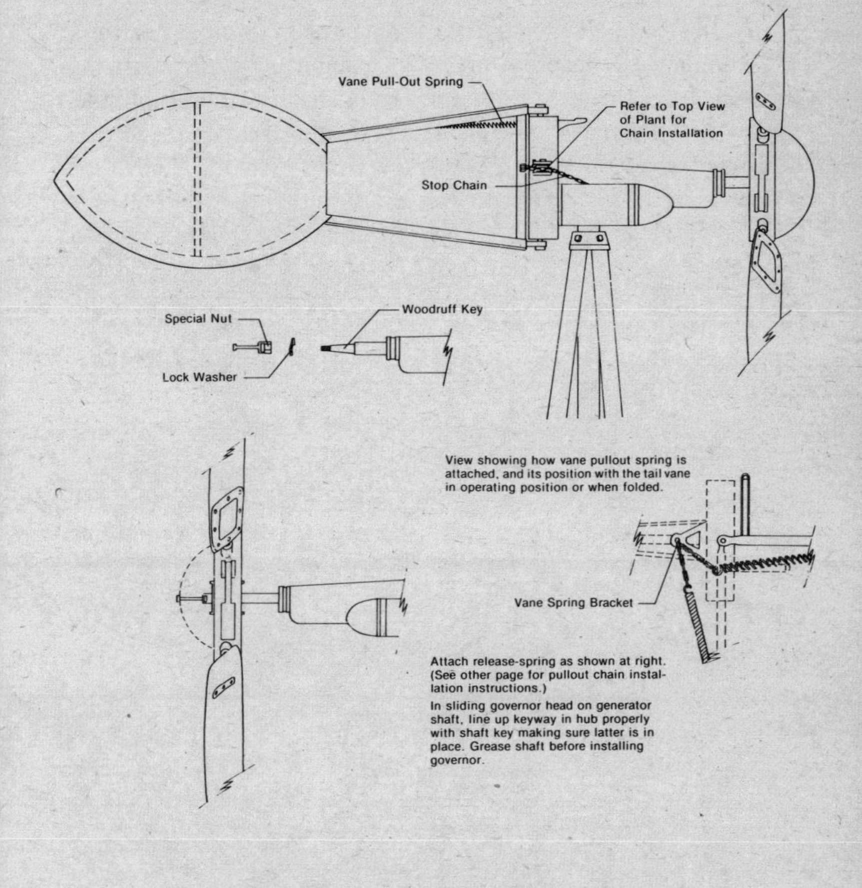

FIGURE 2. Right side view of Jacobs Model 15 (Twin Motor Electric).

and generator characteristics must be matched to produce the best effect.[8] A step-up ratio of 1:10 rotor speed to generator speed should be sufficient to give us the output we need.

Gearing and Lubrication

At this point we should look at some existing wind generators, both commercially manufactured and homemade, to see what we can learn from them. Most of the Jacobs machines, probably the best homestead wind-electric plants ever manufactured, were direct-

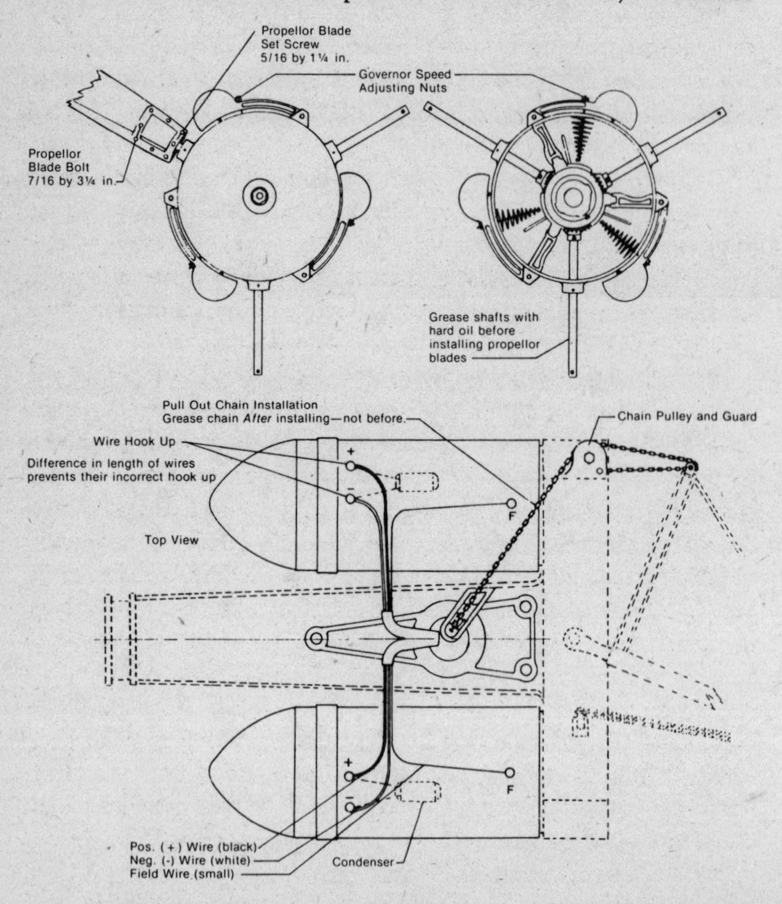

FIGURE 3. Top view of Jacobs Model 15.

drive units—they were designed to put out useable amounts of electricity at the relatively slow speeds of the rotor. To do this required an extremely heavy generator—the 2500-watt Jacobs unit which I own probably weighs in the neighborhood of 500 pounds. (It took four strong men to lift it into my pickup truck, and even then it was not an easy job.)

The old Winchargers, on the other hand, were gear-driven units and much lighter in weight. They didn't produce as many watts as the Jacobs, and because of their more "complicated" design, with gears and consequently higher generator rpm, they didn't last as long.

I have been told that, with the exception of occasional brush replacement, the Jacobs was designed to give a *lifetime* of trouble-free operation. This only stands to reason, given the relationships between friction, speed, and time, a slow-speed generator will last longer than a high speed unit.

Counting the number of gear teeth on a Wincharger I own, I have determined that its gear ratio is about 1:8. These gears are enclosed in a cast-iron housing and turn in a bath of specially formulated light-weight oil, which is changed yearly. The Quirk's machine operates on a similar principle, although the Quirk's people claim that oil changes are necessary at only 5-year intervals.

Now, what about our home-built unit—how are we going to gear it up? The November 1972 issue of *Popular Science* magazine (page 103) has an article and plans for a do-it-yourself wind generator. In looking at these plans we see that a 1:9 gear ratio was achieved by making use of mini-bike sprockets and drive-chain. At first glance, this appears to be the solution to our problem—until we realize that the plans make no provision for an oil bath to lubricate the system. Such a design would necessitate frequent trips up the tower to oil the gears and chain. Unless you enjoy frequent machine maintenance and tower climbing, this system is impractical.

Volume 20 of *The Mother Earth News* (page 32) has another home-built design created by Jim Sencenbaugh of Palo Alto, California. While his wind generator in all other respects looks like a well-designed and efficient unit, we see that the gearing also makes use of mini-bike components, with no apparent oil-bath provision.

The New Alchemy Institute of Woods Hole, Massachusetts, has a wind generator which makes use of an automobile differential mounted on a high tower. The rotor is placed where one wheel hub used to be and turns an automobile alternator at the drive-shaft spline. Again, this appears to be a solution to our gearing problem until we realize that the lowest commonly available auto differential gear ratios are in the neighborhood of 1:4—not nearly enough to get good performance from our aircraft generator.

Of course, we can place a mini-bike gear-up at the drive-shaft spline, but we still haven't solved our problem, because those mini-bike chains will still need lubrication. A gear box for the chains and sprockets could be fabricated, but probably not without some difficulty. On the other hand, a small transmission from a motorcycle or mini-bike might provide adequate gearing here. Bear in mind, however, that the wind energy that is lost in gear friction doesn't generate any electricity, and the more friction we have, the less efficient our system will be. Obviously, then, the fewer gears, chains, and sprockets we use, the better.

One possible way out of our gearing dilemma may be found in the design of the Jacobs Model 15 wind generator. (See Figures 2 and 3.) As was mentioned earlier, *most* Jacobs machines were direct–drive units which had no gearing at all. The Model 15, however, made use of a fly-wheel which turned *two* small 750-watt generators. The rotor turned a short drive-shaft to which the flywheel was attached. The flywheel was encased in a housing to which the two generators were bolted.

If you can imagine an automobile bell housing that has two starter motors bolted to it, one on each side of the engine, you can get an idea of what the Model 15 looks like. In this case, of course, generators replace the starter motors, and instead of an engine with its crankshaft, there is a short driveshaft connecting the rotor to the flywheel. If this sounds confusing, look at Figure 4—the device is amazingly simple.

In addition to solving our gearing and lubrication problems (the housing contains an oil bath), this design gives us an added bonus —*two* generators! If we welded up an oiltight housing, obtained a suitable flywheel from a junked auto engine, geared it properly to *two* of our aircraft generators (suitable gears could be machined or scrounged from auto parts), we would have a wind generator capable of producing 4800 watts of power! (See Figures 4 and 5.) The closest thing you can buy commercially that puts out this much power is the Elektro WV 35 GT 4,000-watt machine, which costs more than $6,000.

Now, I have never constructed such a generator myself, but the Jacobs Company used to make them, and the principle is a valid one. A housing could be fabricated from ¼–inch steel plate. Installation of a "rear main" bearing would be the most complicated part of the job, but anyone reasonably proficient with an arc welder should be able to do it.

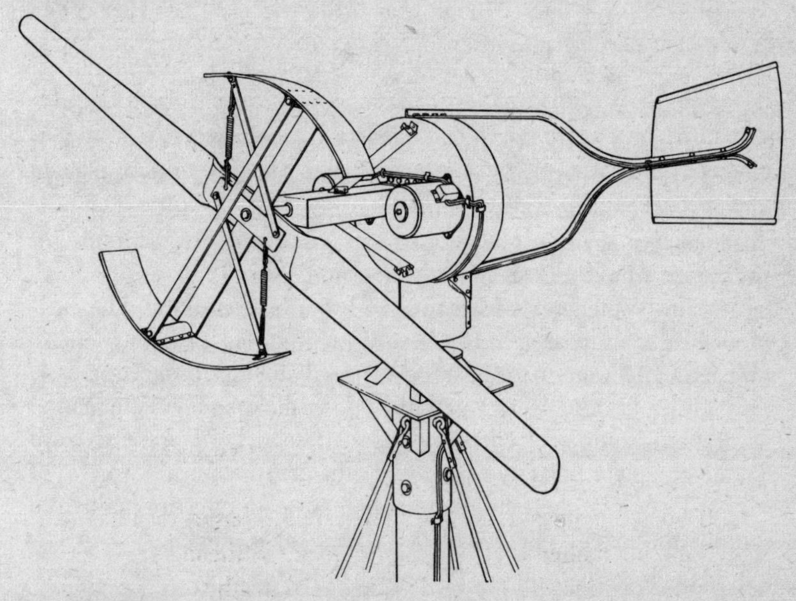

FIGURE 4. Home-built generator based on Jacobs Model 15, capable of producing about 4,800 watts.

The shaft from the rotor is supported by two pillow blocks on a brace, which could be made out of 4- or 6-inch channel iron. There is nothing in the design beyond the skills of any moderately advanced backyard mechanic. The cost of construction should be very little, if any, more than some of the other designs available at this time.

Our wind generator, in theory at least, has overcome the two main drawbacks of home-built wind electric plants: low wattage and complicated gearing to gain the necessary rpm required by generators designed for automotive or aircraft applications. Our task is not yet completed—there are many other factors which must be taken into consideration in constructing a highly efficient unit.

The Rotor

The propeller, or more accurately, the rotor (propellers are devices used to "propel" aircraft), is in some respects the heart of the whole system. If our rotor is not designed and built so as to take maximum advantage of the power in the wind, we can't expect to get much performance from our generator. Again, we can look at other wind generators to get ideas for our home-built unit.

The Jacobs machines had three wooden rotor blades which, when turning in the wind, gave a total rotor diameter of 15 feet. It has been calculated that, when turning at 225 rpm, the centrifugal force on the entire rotor is something like 1100 pounds. Obviously, balancing is of utmost importance for any rotor system we use, or vibration would soon tear the whole unit apart.[9]

The Winchargers used two- and four-bladed rotors. The early four-bladed units and, to the best of my knowledge, all of the two-bladed units, were made of wood. Rotor diameter on the large units was either 12 or 13 feet, depending on the model. (The Wincharger Company manufactured many different models—I presently own three complete units and have parts for many more, yet no two of them are exactly alike, although parts are often interchangeable from model to model.) The two-bladed rotors are essentially a 12-foot 2-×-6 plank which has been worked into the proper aerodynamic shape. Any careful carpenter, proficient with simple wood-working tools, should be able to duplicate such a rotor with ease.

A more sophisticated rotor design, making use of an expandable paper product called Hexcel, which is covered with fiberglass, is described on page 105 of the November 1972 issue of *Popular Science* magazine. This is a three-bladed rotor of the Jacobs type. An experimental two-bladed "sailwing" design is shown on page 71 of the same issue. Complete diagrams for building your own two-bladed wooden rotor originally appeared in *Alternative Sources of Energy* magazine, #8. Plans for Jim Sencenbaugh's complete home-built wind generator (including three-bladed rotor), are de-

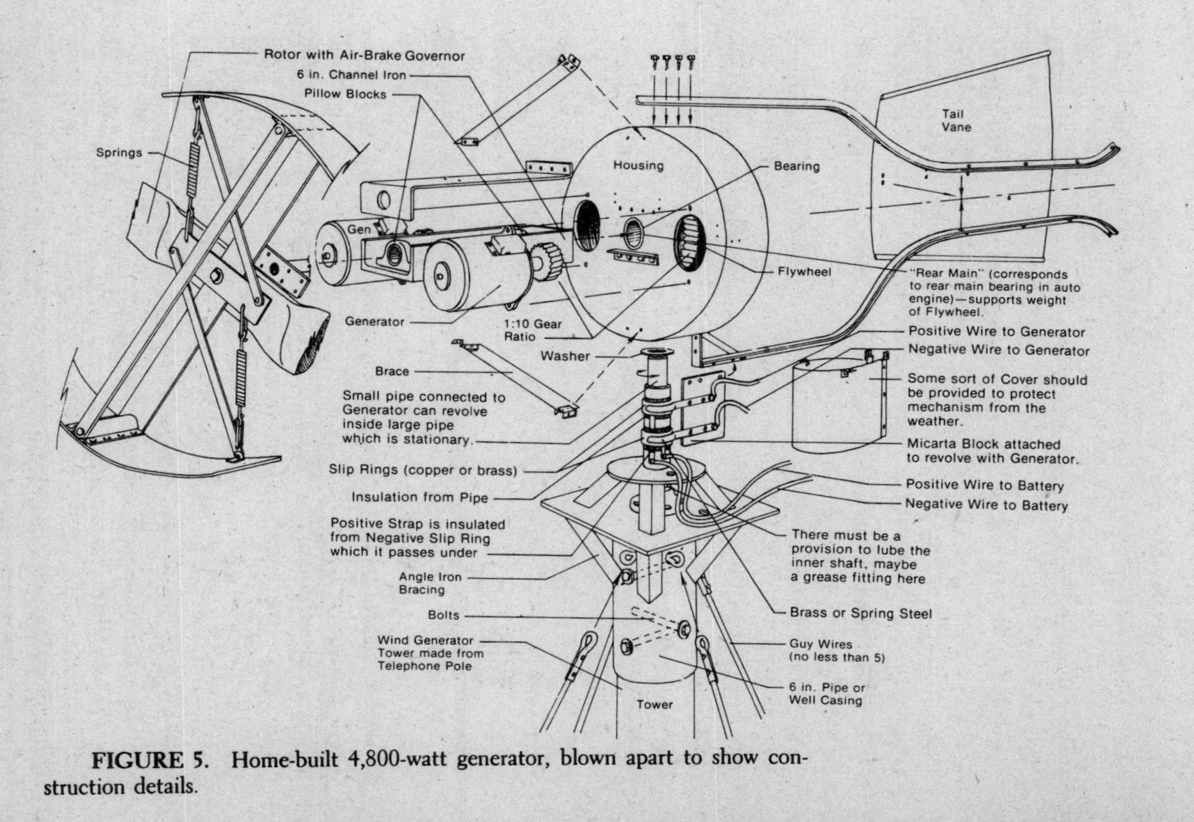

FIGURE 5. Home-built 4,800-watt generator, blown apart to show construction details.

scribed in Volume 20 of *The Mother Earth News,* and are available from the author.

There seem to be some differences of opinion about the ideal number of blades a rotor should have. E. W. Golding, in his definitive book, *The Generation of Electricity by Wind Power* (out of print), states that studies made by the War Production Board during World War II found that the best number of blades is two.[10] These studies, however, were concerned with very large generators to be used for the commercial production of electricity. Most of the homestead–sized generators available today, such as the Quirk's machine, make use of a three-bladed rotor.

In an interview with Marcellus Jacobs (the man who designed and manufactured the fine Jacobs wind generator) in Volume 24 of *The Mother Earth News,* the advantages of the three-bladed rotor are clearly brought out.

For the purposes of our home-built wind generator, however, a two-bladed rotor has two advantages which we should consider: First, a 12-foot wooden rotor, like the Winchargers used, is easy to make; and second, such a rotor design lends itself to being coupled with an airbrake governor, which is also easy to make.

A governing device is absolutely necessary, of course, to slow down the rotor in high winds to prevent damage to the machine. All of the two-bladed Winchargers made use of an airbrake governor—a device which bolted to the front of the rotor in a position perpendicular to the blades. (See Figure 6.) If you can imagine two giant automobile brake shoes held under spring tension on either end of a steel rod, you get a rough picture of what an airbrake governor looks like. At "normal" wind speeds, the spring tension holds the shoes so that they do not interfere with the speed of the rotor. At high wind speeds, centrifugal force overcomes the spring tension, and the "brake shoes" are pulled outward so that they drag through the air and slow the speed of the rotor. *Popular Mechanics* magazine sells plans for a wind generator which makes use of such a governor. Ask for plan number X796 (see Bibliography).

If we take the two-bladed rotor design as described in *Alternative Sources of Energy* #8, extend its dimensions proportionately to make a 12-foot long rotor (plans describe a 7-footer), and couple with that an airbrake governor (also scaled up to conform to the 12-foot length) as described in the *Popular Mechanics* plans, we will have a rotor and governor unit which should work reasonably well with the two aircraft generators we built in our copy of the Jacobs Model 15.

The Tail

Our wind generator now needs a tail vane—a relatively simple device which keeps the machine oriented so that it is always facing the wind. A tail vane from an old wind pump (commonly called "windmills," despite the fact that they pump water rather than grind grain) can easily be adapted to fit our wind generator. Both the Jacobs and Wincharger machines had tail vanes which were designed to be movable—that is, by turning a crank at the base of the tower, the vane could be brought from a position perpendicular to the plane of the rotor to an orientation parallel with it. This had the effect of turning the machine "out of the wind," a maneuver carried out when extremely high winds were expected, or when the batteries were fully charged. Virtually all wind pump vanes are designed to operate the same way, and could be adapted to our generator with little trouble. Should a wind pump vane not be readily available, both the *Popular Mechanics* and Sencenbaugh plans describe how to construct one.

Our generator must be free to turn in any direction in order to capture the full power in the wind. (Although many locations have "prevailing winds," it is actually not true that winds *always* come from a single direction.)[11] A bearing could be improvised using the wheel bearings and hub from an automobile axle—the *Popular Science* design utilizes those off a Volkswagen. The older versions of the Wincharger, however, had no "bearings" at all—the supporting shaft consisted of a shaft or "pipe" which fit snugly inside a slightly larger pipe. The outside section had a grease fitting which provided for lubrication (usually at six-month intervals). Visualize one pipe fitting inside another with a film of grease lubricating them and you have the idea. The LeJay Manual describes this sort of "bearing," as do the Sencenbaugh plans. For ease of construction, this method is probably superior to the automobile wheel bearing idea. (Bear in mind that our wind machine will never be required to pivot anywhere near as fast or often as a car wheel turns, and so a highly sophisticated bearing is not mandatory.)

The Commutator

More complicated, but not nearly as complicated as everyone seems to think, is the construction of a slip-ring commutator. This is a device which allows current to be transmitted from the constantly shifting generator to the stationary wires which run down the tower to the batteries. Many plans for home-made wind generators just allow the wires to twist around the tower—a slipshod compromise which calls into question the validity of the rest of the design. Slip–ring commutators are no big deal—both the *Popular Mechanics* and Sencenbaugh plans show how they are constructed.

The Tower

We are now finished with the generator and its components and are ready to consider our tower. Any wind pump tower of suitable height is easily adapted for wind generator use, but if one is not available, an old telephone pole can also be used; just be sure that the pole is guyed with no fewer than five cables. (That way, if one snaps, you still have four cables to hold the tower in place until you can replace the broken one.) Provision must be made at the top of the tower to hold the generator and its pivoting mechanism. This is easily accomplished by shaving the end of the pole down so that it will just accept a 2-foot length of 6-inch steel pipe of at least ¼ inch wall thickness. At least two (four would be better) long bolts hold the steel pipe to the pole at top and bottom. The bolts should be perpendicular to each other. A square piece of ⅜ inch or ½ inch steel plate is welded to the top of the pipe, and braced with four pieces of angle iron welded to the pipe's sides. (See Figures 4 and 5.) On top of this "platform" is attached the generator's pivoting mechanism and slip-ring commutator and to that, of course, is attached the generator.

One very important consideration must be born in mind when erecting the tower. Most of the criticism of the unsatisfactory opera-

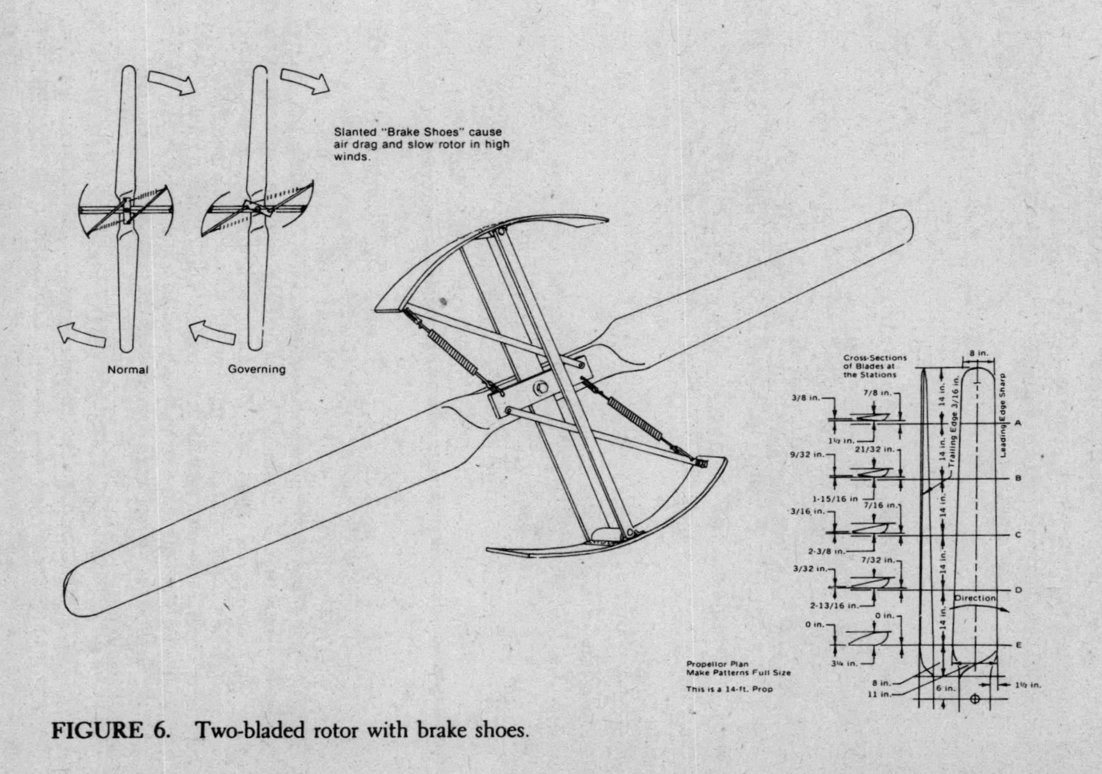

FIGURE 6. Two-bladed rotor with brake shoes.

tion of a wind-driven plant is due to plants being installed in unsuitable positions or on low towers. No plant will work unless it is in a clear air stream and, if a person is not prepared to have a tower of suitable height, it is better not to install the plant. Even if the unit is installed on a clear plain or on top of a hill, it should never be on a tower of less than 30 feet. For all other conditions, a height of at least 40 feet is recommended.[12]

Batteries

Our 24-volt, 4800-watt wind generator is now almost complete. Since the specifications on our aircraft generators state that they have a variable voltage (24 to 32 volts), we will rate the machine at 24 volts, and plan our battery package accordingly—twelve 2-volt, four 6-volt, or two 12-volt cells hooked up in series to provide 24 volts. It is interesting to note that the data plate on my 32-volt Jacobs machine gives its voltage as 40. This is standard procedure, so we don't have to worry that we are harming the 24-volt batteries when our home-built generator is charging at 32 volts. To the best of my knowledge, all DC generators are over-rated a few volts to insure full charge to the batteries (e.g.: "12-volt" automobile generators actually put out 13 or 14 volts).

Because of this relatively "low" voltage (as compared with standard house current of 110 to 120 volts), we will need some fairly

FIGURE 7. Wiring diagram of Jacobs Twin Motor Electric Plant. The field coils are shunt connected: one end of the field coil winding is attached to (+) positive generator brush, then attached to the insulated field terminal at rear end of generator, and from there to the upper collector ring. Then it goes through the third (field) wire to the center terminal "F" terminal, to lower terminal 3, to contact points "A", through frame to position "G", up the right hand wire. From there it runs to the adjustable band on the right–hand resistor coil (maximum charging rate control coil), then down through the resistance wire to the bottom and where wire is attached, returning the circuit to terminal 4 which is attached to the generator negative charging line. This completes the field circuit back to the (−) generator brush.

The "on" and "off" switch in the center of the control panel is open in "on" position. When thrown to the "off" position, it makes a direct circuit from the generator field (center binding post) terminal to the negative charging wire. The purpose of this switch is to give the battery an occasional over-charge by throwing to the "off" position and to make it easy to remove the automatic charging rate control unit from cabinet for replacement or repair. The generator will continue to charge as usual with the switch in the "off" position but, of course, there will be no control of the charging rate. It will be on constant full charging rate.

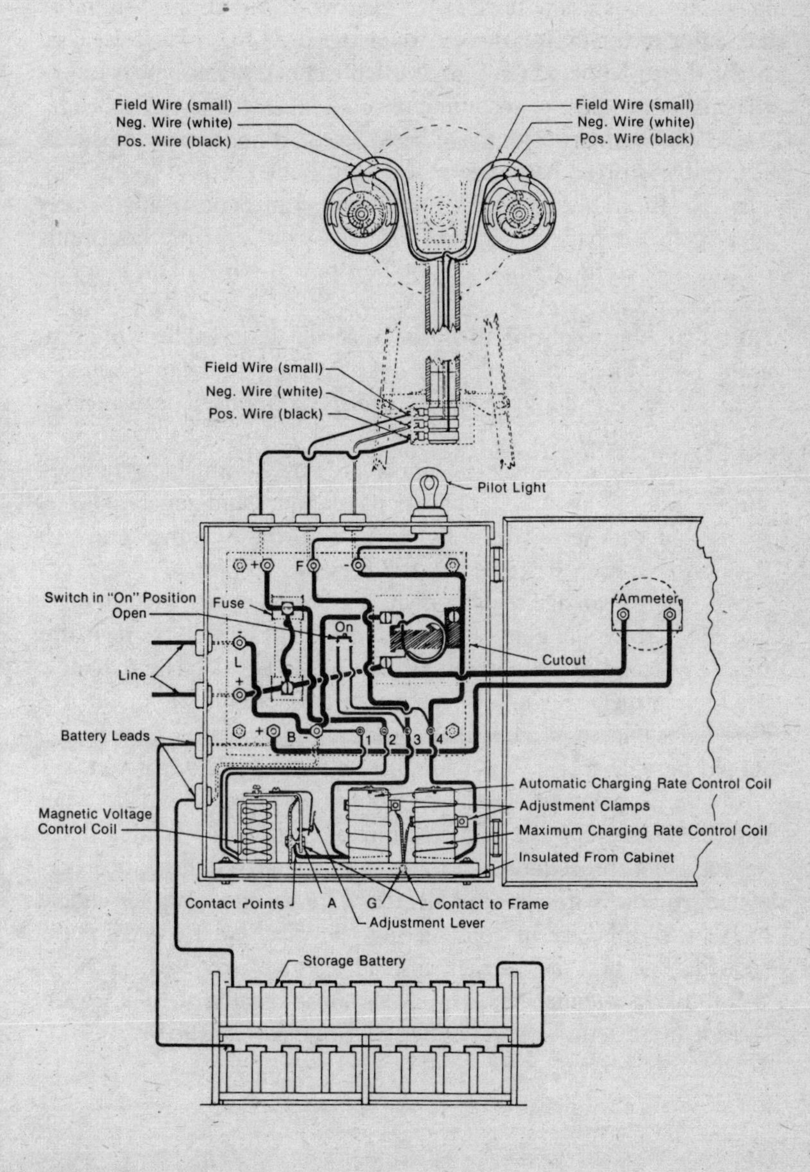
Field Wire (small)
Neg. Wire (white)
Pos. Wire (black)
Field Wire (small)
Neg. Wire (white)
Pos. Wire (black)
Field Wire (small)
Neg. Wire (white)
Pos. Wire (black)
Pilot Light
Switch in "On" Position
Open
Fuse
On
F
L
Line
Ammeter
Cutout
Battery Leads
B
1
2
3
4
Automatic Charging Rate Control Coil
Adjustment Clamps
Maximum Charging Rate Control Coil
Insulated From Cabinet
Magnetic Voltage
Control Coil
Contact Points
A
G
Contact to Frame
Adjustment Lever
Storage Battery

heavy wire to carry the current. The old 32-volt Wincharger and Jacobs manuals recommend No. 6 wire when the distance from the generator to the batteries is less than 100 feet; No. 4 wire was used when the distance was 100 to 200 feet. However, distances longer than 100 feet are *not* recommended.

What sort of batteries are best? There are several options here —2-volt industrial batteries of 180 amp hour rating or higher are available from any battery manufacturer. Jim Sencenbaugh states that "golf cart batteries (Gould PB220, 6-volt 220 amp hour units costing $35 each), are perhaps the best buy in terms of an inexpensive deep cycling unit."[13] Henry Clews' Solar Wind Company sells Australian house lighting batteries especially designed for wind generator use. *Whatever you do, don't use standard automotive batteries —they were never designed for the special conditions under which you will be using them.*

Twenty-four-volt electric motors are readily available as military surplus, but for any other applications for your wind-generated electricity you will need a 24-volt DC to 110-volt AC inverter. Such devices have been described in the previous section.

A 24-volt voltage regulator will need to be constructed to regulate the battery charging rate. I doubt if a surplus voltage regulator from, say, some 24-volt military vehicle would be adequate to handle the high amperage from our two aircraft generators. If you are not knowledgeable about electronic devices, any competent electrician should be able to wire up a suitable voltage regulator for you.

Our do-it-yourself wind generator is now complete. I have not described many of the construction procedures in greater detail because such procedures are more than adequately described elsewhere. Anyone with a modest degree of mechanical aptitude should have little difficulty in constructing the machine I have just described. The best part is that, instead of the ½-kilowatt machines that most do-it-yourselfers create, this one is capable of 4.8 kilowatts —a lot more return on your time and money investment.

2 WATER POWER

Small Water Power Sites

Volunteers in Technical Assistance (VITA)

Flowing water tends to generate automatically a picture of easy, free energy in the eyes of someone who's looking for a source of "homemade" power. Don't be deceived. Harnessing water power is always going to cost *something,* and there are many factors that must be considered before you begin to dam up that babbling stream or rushing river running through your land.

First, check with local ordinances to see if building dams and other structures on waterways is permitted. Then make sure that any alterations you will make will not harm the wildlife or fish in your area or interfere with someone else's use of that same water downstream from you.

Your costs will be relatively low if the head, or the height of the body of water, considered as causing pressure, is relatively high, because a fairly inexpensive turbine can be used. (See Figure 8 for measuring head.) Water power can also be economical where a dam can be built into a small river with a relatively short (less than 100 feet) conduit (penstock) for conducting water to the water wheel. (See Figure 13 for such an installation.) Costs will be fairly high, however, when such a dam and pipeline can provide a head of only 20 feet or less. If you have not had welding experience and will therefore have to buy a turbine or water wheel rather than make it, harnessing water power this way could prove to be a relatively expensive project for you.

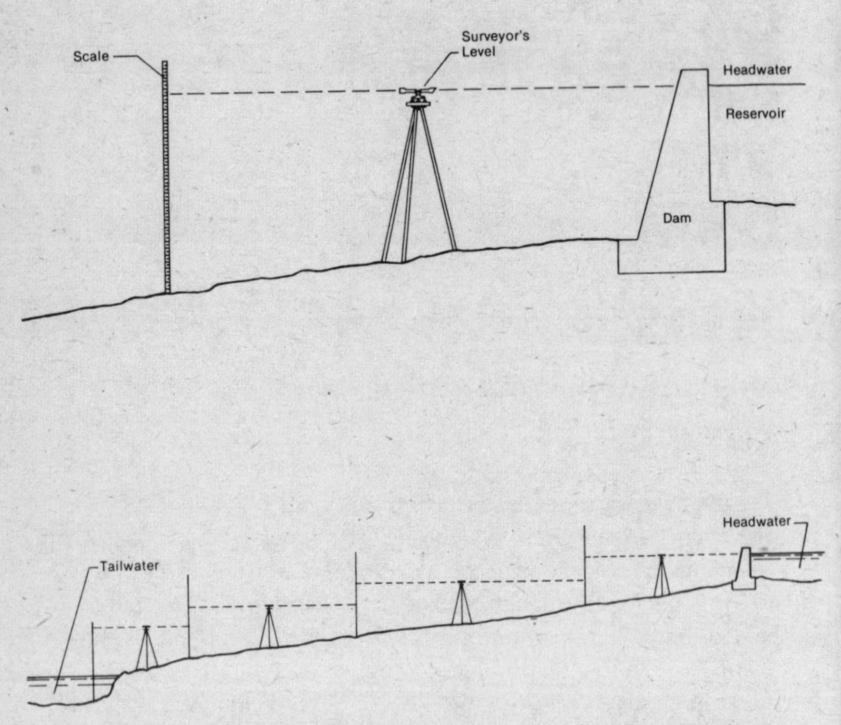

FIGURE 8. Measuring head with surveyor's level (Total gross head = A + B + C + D).

The water power stations described here are best suited for small communities rather than for individual homesteads. If you want to harness an appreciable amount of power from a stream or river it will be necessary to build a station like the ones you'll find here, and they all cost money and require a good amount of work. They will usually prove to be economical only when the work and construction costs can be split among many people.

Getting Started

You'll find that turbine manufacturers will furnish you with a considerable amount of advice and usually outline a drawing of the entire project for you if you supply them with the data listed below.

1. Minimum flow of water available in cubic feet per second (or cubic meters) per second
2. Maximum flow of water available in cubic feet per second (or cubic meters) per second
3. Head or fall of water in feet (or meters)
4. Length of pipeline in feet (or meters) needed to get the required head
5. Describe water condition (clear, muddy, sandy, acid)
6. Describe soil condition (see Table 7)
7. Minimum tailwater elevation in feet (or meters)
8. Approximate area of pond above dam in acres (or square kilometers)
9. Approximate depth of the pond in feet (or meters)
10. Distance from power plant to where electricity will be used in feet (or meters)
11. Approximate distance from dam to power plant
12. Minimum air temperature
13. Maximum air temperature
14. Estimate power to be used
15. A sketch with elevations, or topographical map with site sketched in

Power

Before you begin plans for your water power site, you should determine the amount of power you'll need. For more information on calculating your power requirements, check Appendix I. The required amount of power (gross power) is equal to the useful power plus the losses inherent in any power scheme. It is usually safe to assume that the net or useful power in the case of small power installations will only be half of the available gross power, due to water transmission losses and the turbine and generator efficiencies. Some power is lost when it is transmitted from the generator switchboard to the place of application. The GROSS POWER, the power available from the water, is determined by the following formula:

In English Units: gross power (horsepower) =

$$\frac{\text{minimum water flow (cubic feet/second)} \times \text{gross head (feet)}}{8.8}$$

In Metric Units: gross power (metric horsepower) =

$$\frac{1{,}000}{75}\ \text{flow (cubic meters/second)} \times \text{head (meters)}$$

The NET POWER available at the turbine shaft is:

In English Units: net power =

$$\frac{\text{minimum water flow} \times \text{net head}}{8.8} \times \text{turbine efficiency (English)}$$

In Metric Units: net power =

$$\frac{\text{minimum water flow} \times \text{net head}}{75/1{,}000} \times \text{turbine efficiency (metric)}$$

The NET HEAD is obtained by deducting the energy losses from the gross head (see section, **Measuring Head Losses**). A good assumption for turbine efficiency, when it is not known, is 80 percent.

Measuring Gross Head (either method)

Method No. 1

Equipment:

Surveyor's leveling instrument—consists of a spirit level fastened parallel to a telescopic sight

Scale—use wooden board approximately 12 feet in length

Procedure (note Figure 8):

A surveyor's level on a tripod is placed downstream from the power reservoir dam on which the headwater level is marked. After taking a reading, the level is turned 180° in a horizontal circle. The scale is placed downstream from it at a suitable distance and a second reading is taken. This process is repeated until the tailwater level is reached.

Method No. 2

This method is fully reliable, but is more tedious than Method No. 1 and need only be used when a surveyor's level is not available.

Equipment:

Scale
Board
Ordinary carpenter's level

Procedure (note Figure 9):

Place board horizontally at headwater level and place level on top of it for accurate leveling. At the downstream end of the horizontal board, the distance to a wooden plug set into the ground is measured with a scale. The process is repeated until the tailwater level is reached.

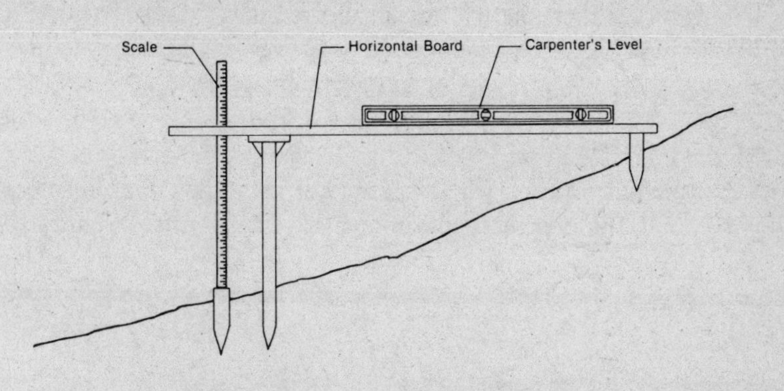

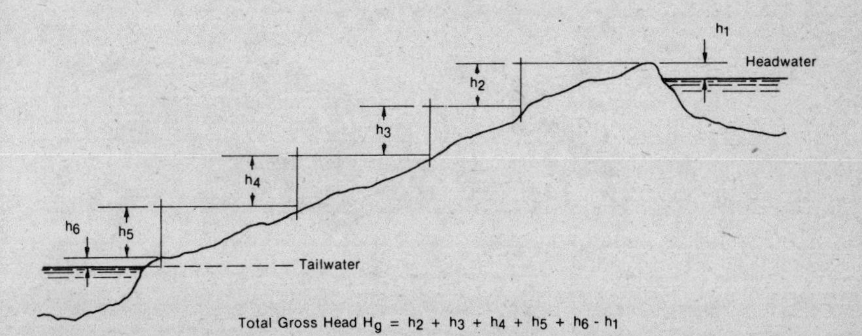

FIGURE 9. Measuring head with carpenter's level.

Measuring Flow Rate

For power purposes, measurements should take place at the season of lowest flow (usually in autumn) in order to guarantee full power at all times. Investigate the stream flow history to ascertain that the minimum required flow is that which has occurred for as many years as it is possible to determine. If there have been years of drought in which flow rate was reduced below the minimum required, other streams or sources of power may offer a better solution.

Method No. 1

For small streams with a capacity of less than one cubic foot per second, build a temporary dam in the stream, or use a "swimming hole" created by a natural dam. Channel the water into a pipe and catch it in a bucket of known capacity. Determine the stream flow by measuring the time it takes to fill the bucket.

$$\text{stream flow (cubic feet per second)} = \frac{\text{volume of bucket (cubic feet)}}{\text{filling time (seconds)}}$$

Method No. 2

For medium streams with a capacity of more than one cubic foot per second, the weir method can be used. The weir (see Figures 10

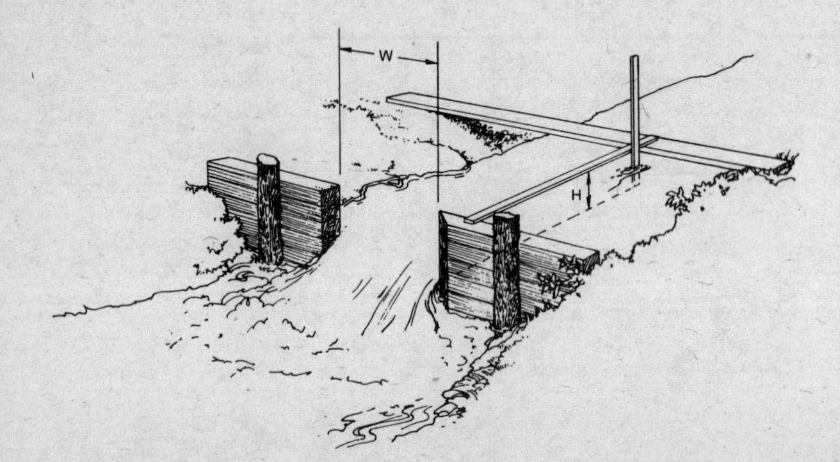

FIGURE 10. The weir method of measuring flow.

and 11) is made from boards, logs, or scrap lumber. Cut a rectangular opening in the center. Seal the seams of the boards and the sides built into the banks with clay or sod to prevent leakage. Saw the edges of the opening on a slant to produce sharp edges on the upstream side. A small pond is formed upstream from the weir.

When there is no leakage and all water is flowing through the weir opening, (1) place a board across the stream and (2) place another narrow board level (use a carpenter's level) and perpendicular to the first. Measure the depth of the water above the bottom edge of the weir with the help of a stick on which a scale has been marked. Determine the flow from Table 6.

TABLE 6. Flow Value (cubic feet per second)

	WEIR WIDTH						
OVERFLOW HEIGHT	*3 feet*	*4 feet*	*5 feet*	*6 feet*	*7 feet*	*8 feet*	*9 feet*
1.0 inch	.24	.32	.40	.48	.56	.64	.72
2 inches	.67	.89	1.06	1.34	1.56	1.8	2.0
4 inches	1.9	2.5	3.2	3.8	4.5	5.0	5.7
6 inches	3.5	4.7	5.9	7.0	8.2	9.4	10.5
8 inches	5.4	7.3	9.0	10.8	12.4	14.6	16.2
10 inches	7.6	10.0	12.7	15.2	17.7	20.0	22.8
12 inches	10.0	13.3	16.7	20.0	23.3	26.6	30.0

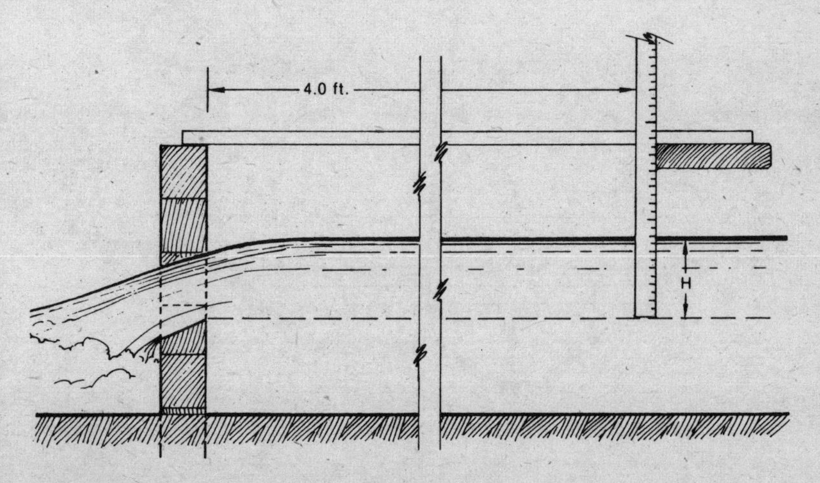

FIGURE 11. Measuring flow with a weir.

Method No. 3

The float method (Figure 12) is used for larger streams. Although it is not as accurate as the previous two methods, it is adequate for practical purposes. Choose a point in the stream where the bed is smooth and the cross section is fairly uniform for a length of at least 30 feet. Measure water velocity by throwing pieces of wood into the water and measuring the time of travel between two fixed points, 30 feet or more apart. Erect posts on each bank at these points (four posts in all).

Connect the two upstream posts by a level wire rope (use a carpenter's level). Follow the same procedure with the downstream posts. Divide the stream into equal sections along the wires and measure the water depth for each section. In this way, the cross-sectional area of the stream is determined. Use the following formula to calculate the flow: stream flow (cubic feet per second) = average cross-sectional flow area (square feet) × velocity (feet per second).

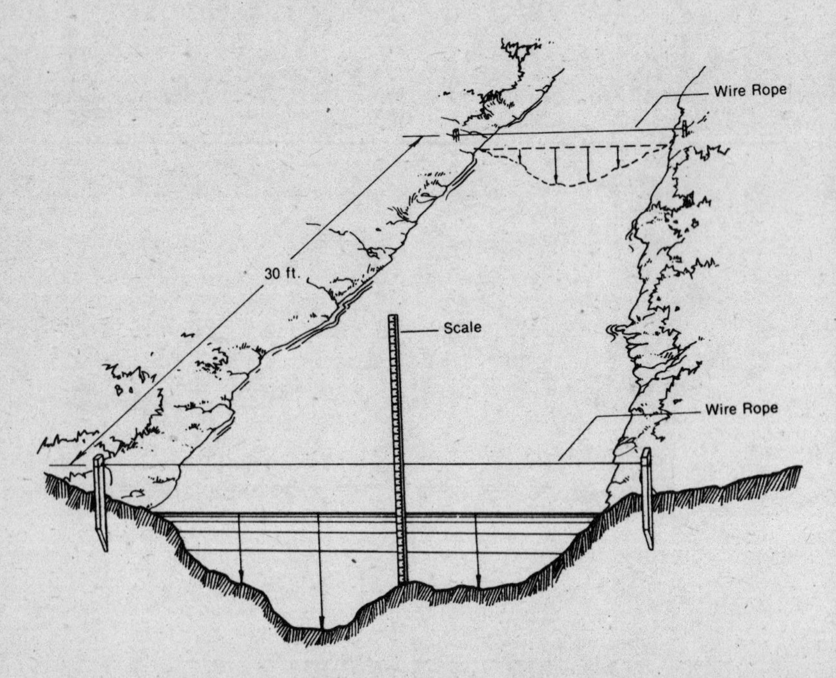

FIGURE 12. The float method of measuring flow.

Measuring Head Losses

As noted in the section, **Power**, net power is a function of the net head. The net head is the gross head less the head losses. Figure 13 shows a typical small water power installation. The head losses are the open-channel losses plus the friction loss from flow through the penstock.

Open Channel Head Losses

The headrace and the tailrace in Figure 14 are open channels for transporting water at low velocities. The walls of channels made of

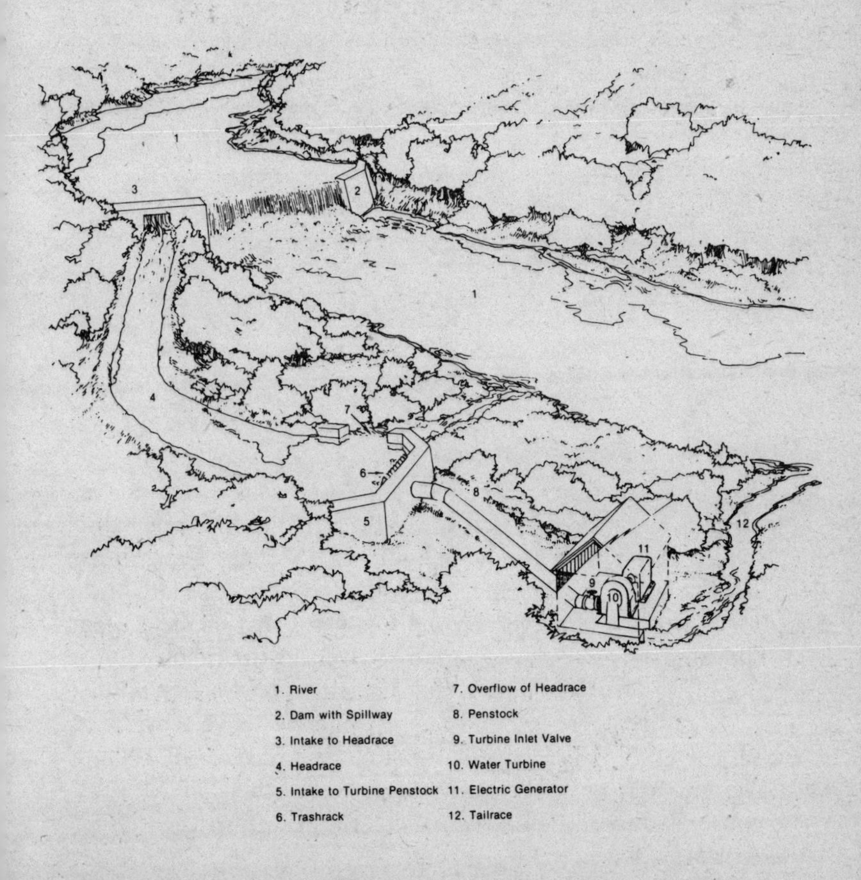

FIGURE 13. A typical installation for a low-output water power plant.

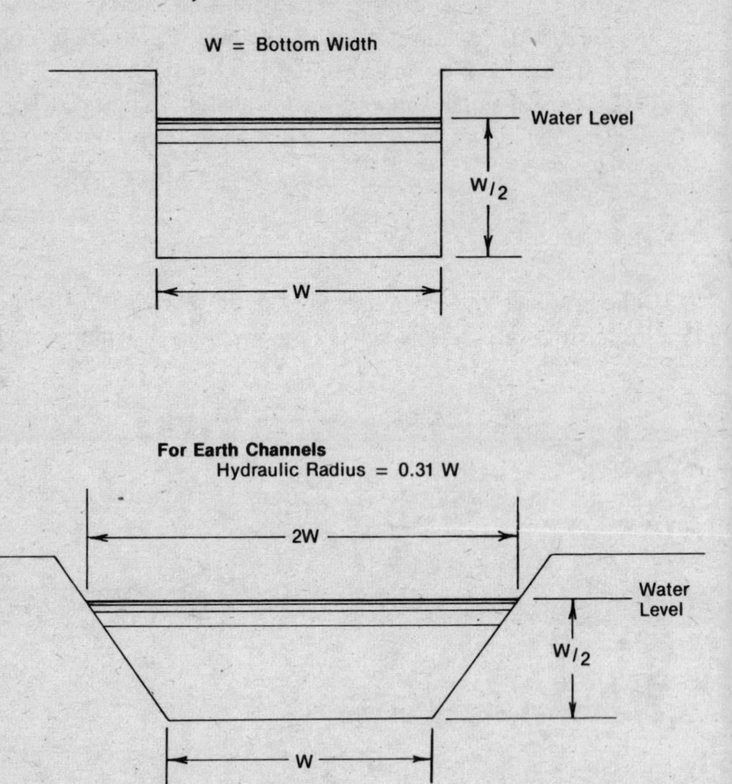

FIGURE 14. Building water channels.

timber, masonry, concrete, or rock should be constructed perpendicularly. Design them so that the water level height is one half of the width. Earth walls should be built at a 45° angle. Design them so that the water level height is one half of the channel width at the bottom. At the water level the width is twice that of the bottom.

The head loss in open channels is given in the nomograph in Figure 15. The friction effect of the material of construction is called "n." Various values of "n" and the maximum water velocity, below which the walls of a channel will not erode, are given in Table 7.

The *hydraulic radius* is equal to a quarter of the channel width, except for earth-walled channels where it is 0.31 times the width at the bottom.

TABLE 7. Friction Effects of Channel Construction Materials

MATERIAL OF CHANNEL WALL	MAXIMUM ALLOWABLE WATER VELOCITY (FEET/SECOND)	VALUE OF "N"
Fine grained sand	0.6	0.030
Coarse sand	1.2	0.030
Small stones	2.4	0.030
Coarse stones	4.0	0.030
Rock	25.0	(Smooth) 0.033 (Jagged) 0.045
Concrete with sandy water	10.0	0.016
Concrete with clean water	20.0	0.016
Sandy loam, 40% clay	1.8	0.030
Loamy soil, 65% clay	3.0	0.030
Clay loam, 85% clay	4.8	0.030
Soil loam, 95% clay	6.2	0.030
100% clay	7.3	0.030
Wood		0.015
Earth bottom with rubble sides		0.033

To use the nomograph, draw a straight line from the value of "n" through the flow velocity to the reference line. The point on the reference line is connected to the hydraulic radius, and this line is extended to the head-loss scale which also determines the required slope of the channel.

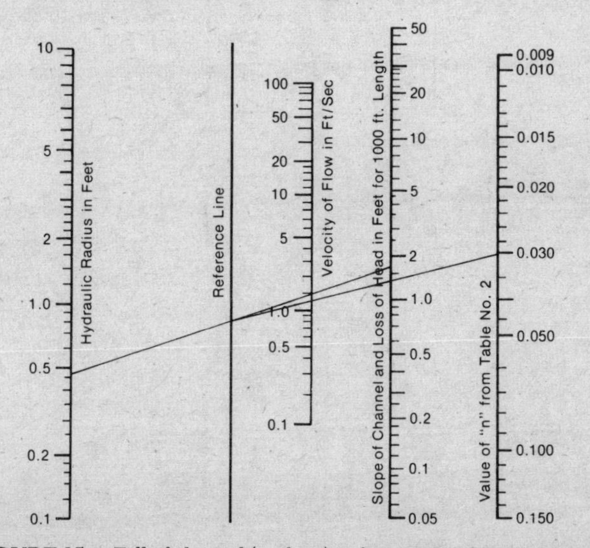

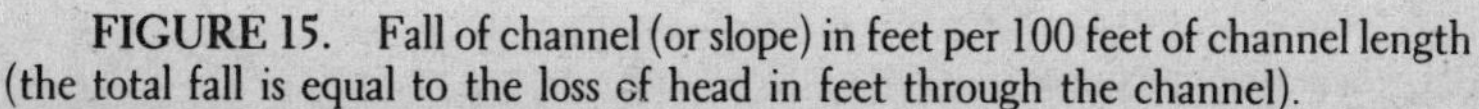

FIGURE 15. Fall of channel (or slope) in feet per 100 feet of channel length (the total fall is equal to the loss of head in feet through the channel).

PIPE HEAD LOSS AND PENSTOCK INTAKE

The *trashrack* in Figure 16 is a weldment consisting of a number of vertical bars held together by an angle at the top and a bar at the bottom. The vertical bars must be spaced in such a way that the teeth of a rake can penetrate the rack for removing leaves, grass, and trash which might clog up the intake. Such a trashrack can easily be manufactured in the field or in a small welding shop.

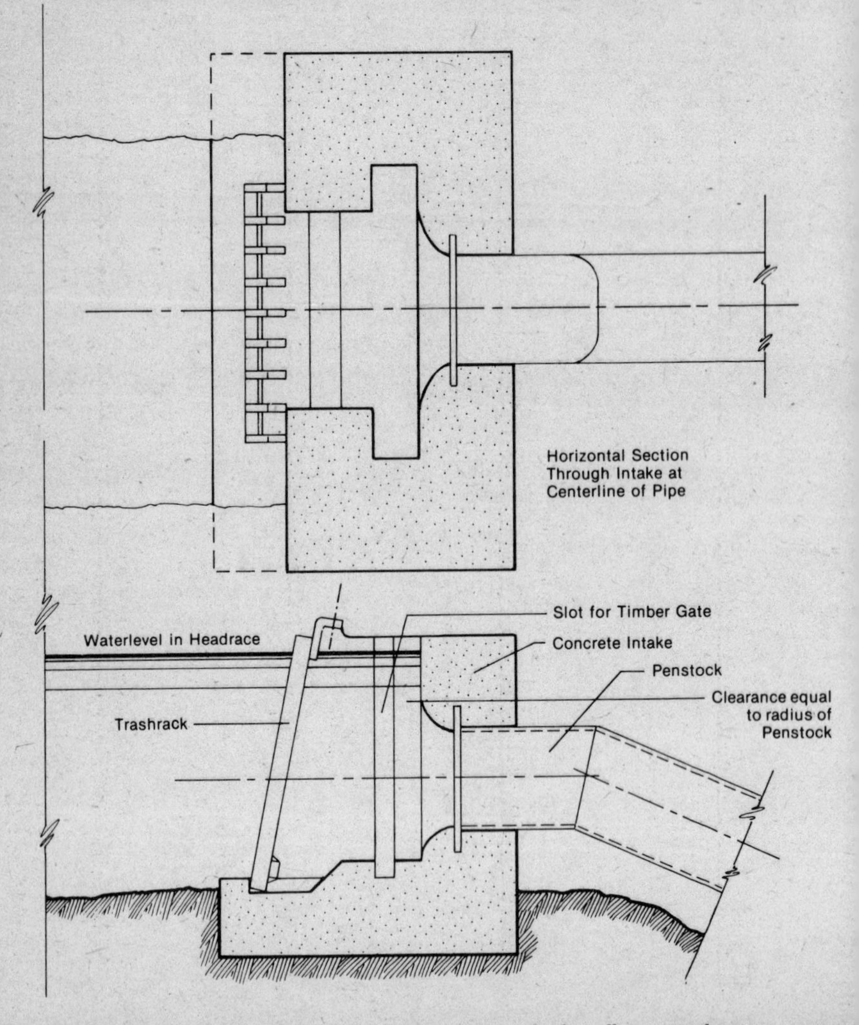

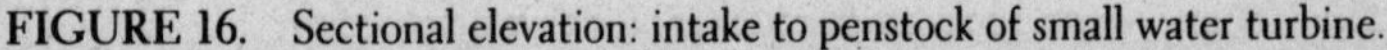

FIGURE 16. Sectional elevation: intake to penstock of small water turbine.

Downstream from the trashrack, a slot is provided in the concrete into which a timber gate can be inserted for shutting off the flow of water to the turbine.

The penstock can be constructed from commercial pipe. The pipe must be large enough to keep the head loss small. From the nomograph (Figure 17) the required pipe size is determined. A straight line drawn through the water velocity and flow rate scales gives the

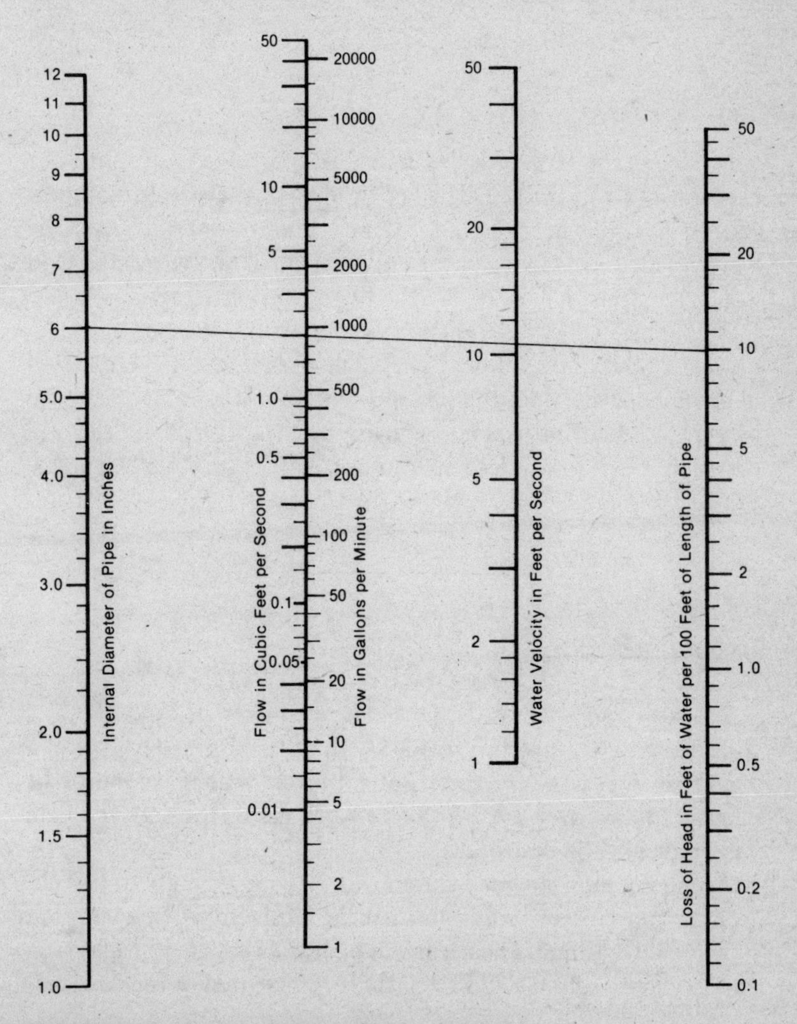

FIGURE 17. Nomograph for head loss in steel pipes.

required pipe size and pipe head loss. Head loss is given for a 100-foot pipe length. For longer or shorter penstocks, the actual head loss is the head loss from the chart multiplied by the actual length divided by 100. If commercial pipe is too expensive, it is possible to make pipe from native material; for example, concrete and ceramic pipe or hollowed logs. The choice of pipe material and the method of making the pipe depend on the cost and availability of labor and the availability of material.

Small Dams

In most cases a dam is necessary to direct the water into the channel intake or to get a higher head than the stream naturally affords. A dam is not required if there is enough water to cover the intake of a pipe or channel at the head of the stream where the dam would be placed.

A dam may be made of earth, wood, concrete, or stone. In building any kind of dam, all mud, vegetable matter, and loose material must be removed from the bed of the stream where the dam is to be placed. This usually is not difficult since most small streams will cut their beds down close to bedrock, hard clay, or other stable formation.

EARTH DAMS

An earth dam may be desirable where concrete is expensive and timber scarce. It must be provided with a separate spillway of sufficient size to carry off excess water because water can never be allowed to flow over the crest of an earth dam. If it does the dam will erode and be destroyed.

A spillway must be lined with boards or with concrete to prevent seepage and erosion. Still water is held satisfactorily by earth, but moving water is not. The earth will be worn away by it. Figures 18 and 19 show a spillway and an earth dam. The crest of the dam may

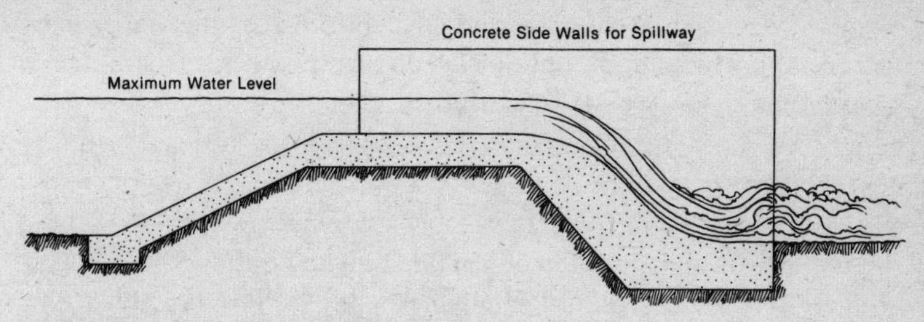

FIGURE 18. Concrete spillway for earth-fill dam.

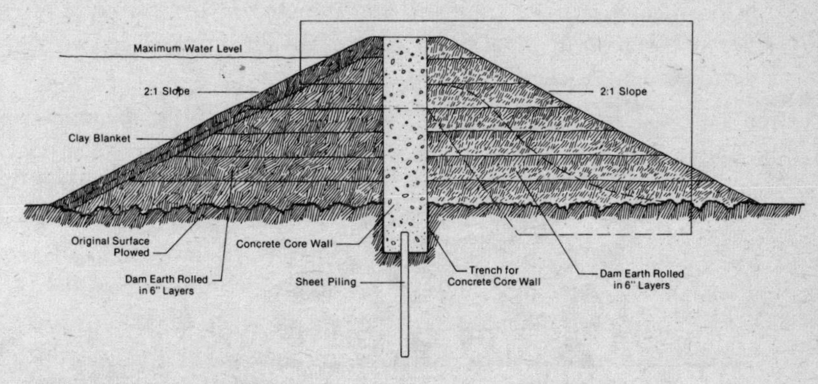

FIGURE 19. Earth-fill dam.

be just wide enough for a footpath or may be wide enough for a roadway, with a bridge placed across the spillway.

The greatest difficulty in earth dam construction occurs in places where the dam rests on solid rock. It is hard to keep the water from seeping between the dam and the earth and finally undermining the dam. One way to prevent seepage is to blast and clean out a series of ditches in the rock, with each ditch about a foot deep and 2 feet wide extending under the length of the dam. Each ditch should be filled with 3 or 4 inches of wet clay compacted by stamping it. More layers of wet clay can then be added and the compacting process repeated each time until the clay is several inches higher than bedrock.

The upstream half of the dam, as shown in Figure 19, should be of clay or heavy clay soil, which compacts well and is impervious to water. The downstream side should consist of lighter and more porous soil, which drains out quickly and thus makes the dam more stable than if it were made entirely of clay.

CRIB DAMS

The crib dam is very economical in timber country, as it requires only rough tree trunks, cut planking, and stones. Four to 6-inch tree trunks are placed 2 to 3 feet apart and spiked to others placed across them at right angles. Stones fill the spaces between timbers. The upstream side (face) of the dam, and sometimes the downstream side, are covered with planks. (See Figure 20.) The face is sealed with clay to prevent leakage.

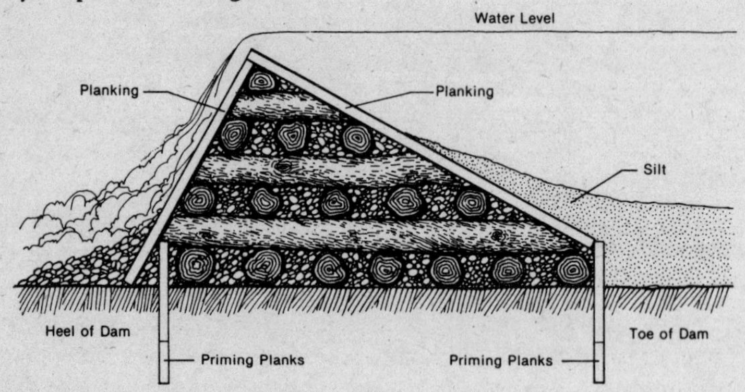

FIGURE 20. Crib dam with downstream planking.

Downstream planks are used as an apron to guide the water which overflows the dam back into the stream bed. The dam itself serves as a spillway in this case. The water coming over the apron falls rapidly, and it is necessary to line the bed below with stones in order to prevent erosion. A section of crib dam without downstream planking is illustrated in Figure 21. The apron consists of a series of steps for slowing the water gradually.

Crib dams, as well as other types, must be embedded well into the embankments and packed with impervious material, such as clay or heavy earth and stones, in order to anchor them and to prevent leakage. At the heel, as well as at the toe of crib dams, longitudinal rows of planks are driven into the stream bed. These are priming planks which anchor the dam and prevent water from seeping under

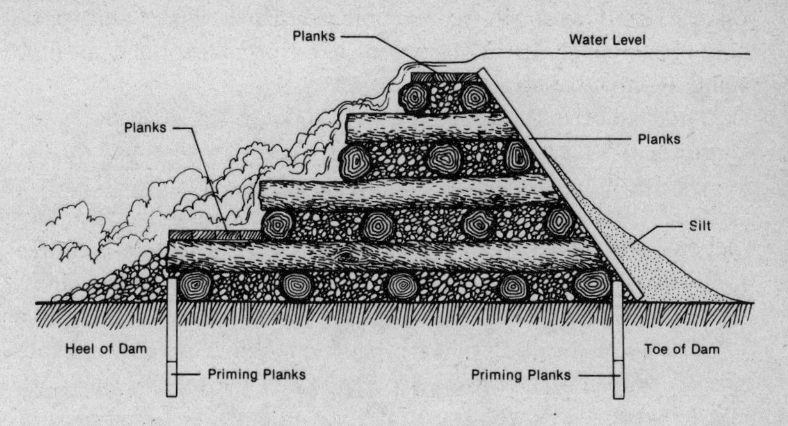

FIGURE 21. Crib dam without downstream planking.

it. If the dam rests on rock, priming planks cannot and need not be used, but where the dam does not rest on rock, they make it more stable and watertight.

These priming planks should be driven as deep as possible and then spiked to the timber of the crib dam. The lower ends of the priming planks are pointed as shown in Figure 22, and they must be placed one after the other as shown. Thus each successive plank is forced, by the act of driving it, closer against the preceding plank, resulting in a solid wall.

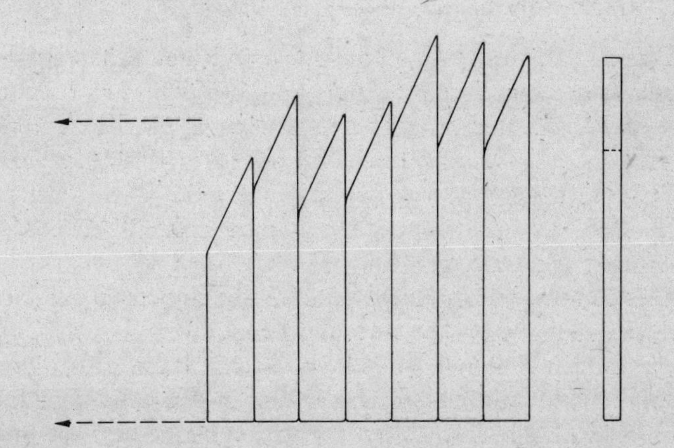

FIGURE 22. Priming planks.

Any rough lumber may be used. Chestnut and oak are considered the best material. The lumber must be free from sap, and its size should be approximately 2-by-6 inches.

In order to drive the priming planks and also the sheet piling of Figure 19, considerable force may be required. A simple pile driver as shown in Figure 23 will serve the purpose.

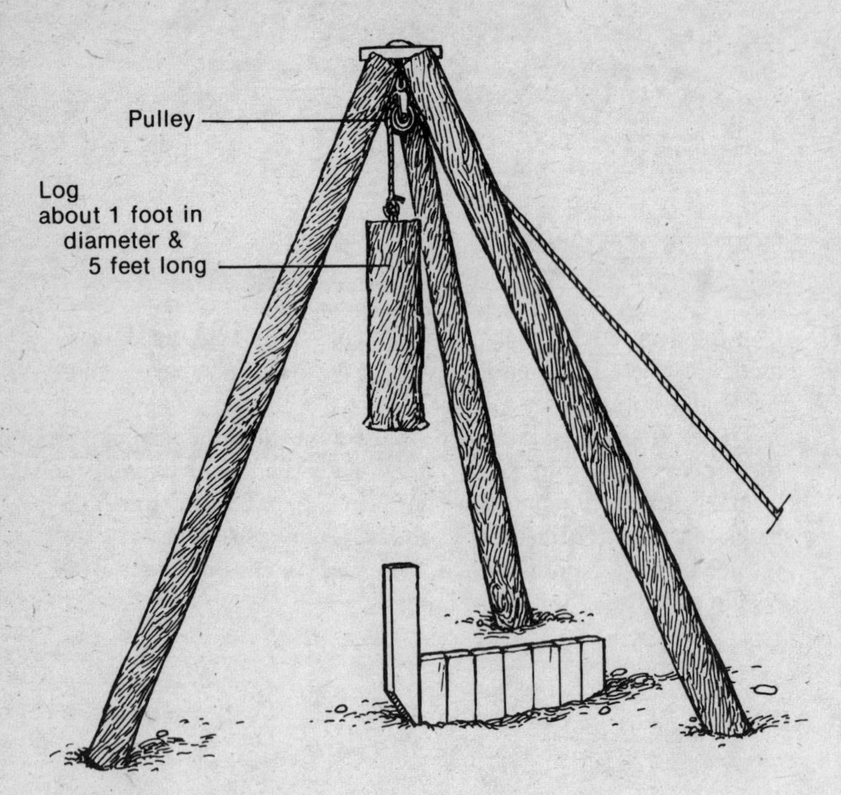

FIGURE 23. Pile driver.

CONCRETE AND MASONRY DAMS

Concrete and masonry dams more than 12 feet high should not be built without the advice of a competent engineer with experience in this special field. Dams of less height require knowledge of the soil condition and bearing capacity, as well as of the structure itself.

Figure 24 shows a stone dam which also serves as a spillway. Such a dam can be as high as 10 feet. It is made of rough stones, but the layers should be bound by concrete. The dam must be built down to a solid and permanent footing to prevent leakage and shifting. The base of the dam should have the same dimension as its height to give it stability.

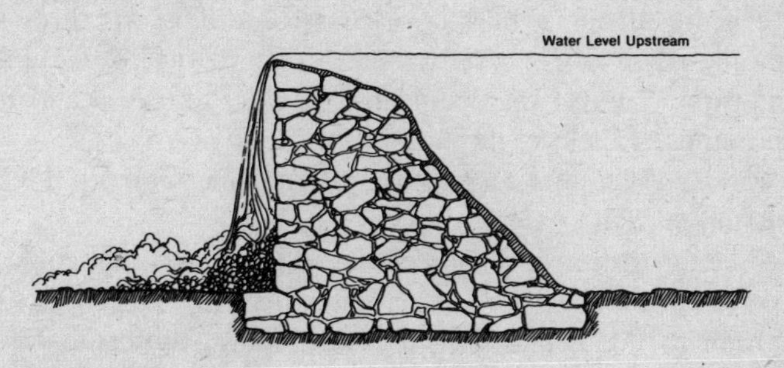

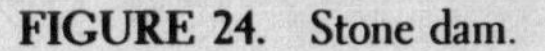
FIGURE 24. Stone dam.

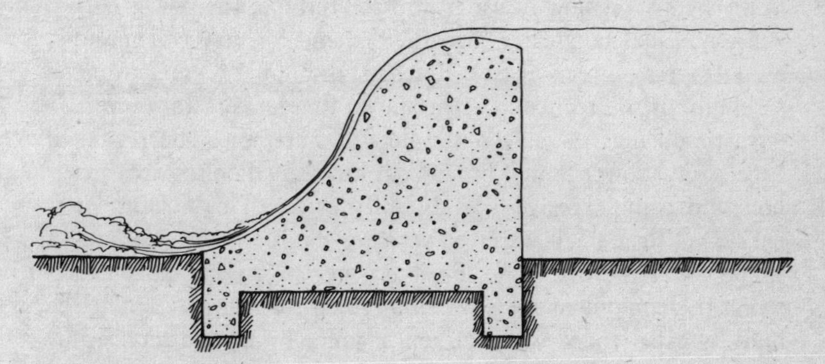

FIGURE 25. Small concrete dam.

Small concrete dams (Figure 25) should have the thickness of their bases 50 percent greater than their height. The apron is designed to turn the flow slightly upwards to dissipate the energy of the water and protect the downstream bed from eroding.

Water Turbines

The manufacturers of hydraulic turbines for small plants can usually give you a quote on a complete packaged unit, including the generator, governor, and switch gear. Water turbines for small power developments may be purchased (see Table 8) or made in the field, if a small machine and weld shop are available.

A centrifugal pump may be used as a turbine wherever it is technically possible. Its cost is about one-third the cost of an hydraulic turbine, but it may be poor economics to use a centrifugal pump because it is less efficient than a turbine.

A water power unit can produce either direct current (DC) or alternating current (AC) electricity.

Two factors to consider in deciding whether to install an AC or DC power unit are (1) the cost of regulating the flow of water into the turbine for AC and (2) the cost of converting motors to use DC electricity.

FLOW REGULATION

The demand for power will vary from time to time during the day. With a constant flow of water into the turbine, the power output will sometimes be greater than the demand for power. Therefore, either excess power must be stored or the flow of water into the turbine must be regulated according to the demand for power.

In producing AC power, the flow of water must be regulated because AC power cannot be stored. Flow regulation requires governors and complex valve-type shut-off devices. This equipment is expensive; in a small water power site, the regulating equipment would cost more than a turbine and generator combined. Furthermore, the equipment for any turbine used for AC power must be built by experienced water-turbine manufacturers and serviced by competent consulting engineers.

The flow of water to a DC power-producing turbine, however, does not have to be regulated. Excess power can be stored in a storage battery. DC generators and storage batteries are low in cost because they are mass-produced.

To summarize: In producing AC power, the flow of water into the

TABLE 8. Small Hydraulic Turbines

	TYPES		
	Impulse or Pelton	*Michell or Banki*	*Centrifugal pump used as turbine*
Head Range	50 to 1000	3 to 650	
Flow Range (cubic feet per second)	0.1 to 10	0.5 to 250	Available for any desired condition
Application	high head	medium head	
Power (horsepower)	1 to 500	1 to 1000	
Cost per kilowatt	low	low	low
Manufacturers	James Leffel & Co. Springfield, OH 45501	Ossberger-Turbinenfabrik 8832 Weissenburg Bayern Bavaria, Germany	
	Drees & Co. Werl, Germany	Can be do-it-yourself project if small weld and machine shops are available	
	Officine Buhler Taverne, Switzerland		

turbine must be regulated; this requires costly and complex equipment. In producing DC power, regulation is not necessary, but storage batteries must be used.

Converting Motors for DC Power

DC power is just as good as AC for producing electric light and heat, but for electrical appliances—from farm machinery to household appliances—the use of DC power can involve some expense. When such appliances have AC motors, DC motors must be installed. The cost of these alterations must be weighed against the cost of flow regulation needed for producing AC power.

Impulse Turbines

Impulse turbines are used for high heads and low flow rates. They are the most economical turbines because the high head gives them high speed, and their size and weight per horsepower is small. Construction costs of intake and power house are also small. A very simplified version is shown in Figures 26 and 27.

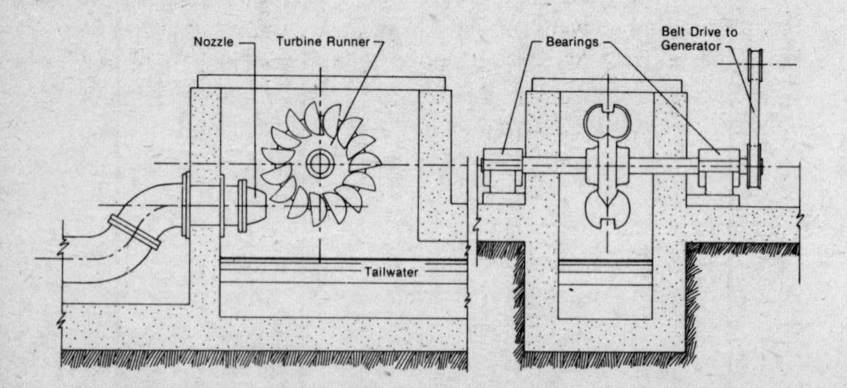

FIGURE 26. Small impulse turbine in concrete housing.

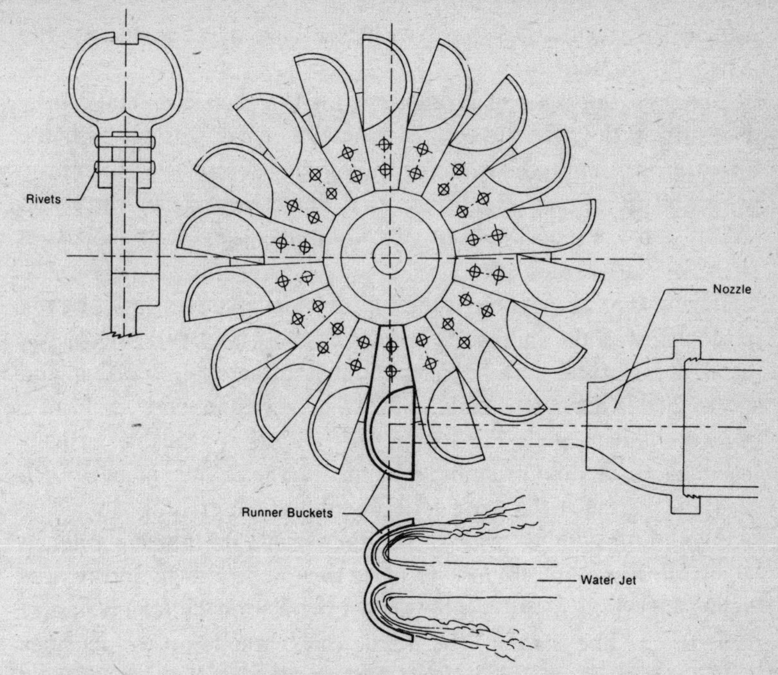

FIGURE 27. Small impulse-turbine runner.

Michell (Banki) Turbine

The Michell (or Banki) turbine is simple in construction and may be the only type of water turbine which can be locally built. The two main parts of the Michell turbine are the runner and the nozzle. Both are welded from plate steel and require some machining. Welding equipment and a small machine shop like those often used to repair farm machinery and automotive parts are all that are necessary.

Figures 28 and 29 show the arrangement of a turbine of this type for low-head use without control. This installation drives a DC generator with a belt drive. Because the construction can be a do-it-yourself project, formulas and design details are given for a runner with a 12-inch outside diameter. This is the smallest size that is easy to fabricate and weld. It has a wide range of applications for all small power developments with head and flow suitable for the Michell turbine.

Different heads result in different rotational speeds. The proper belt-drive ratio gives the correct generator speed. Various amounts of water determine the width of the nozzle (B_1, Figure 29) and the width of the runner (B_2, Figure 29). These widths may vary from 2 inches to 14 inches. No other turbine is adaptable to as large a range of flow as this one is.

The water passes through the runner twice in a narrow jet before discharge into the tailrace. The runner consists of two side plates, each ¼ inch thick with hubs for the shaft attached by welding, and 20 to 24 blades. Each blade is 0.237 inches thick and cut from a 4-inch standard pipe. Steel pipe of this type is available virtually everywhere. A pipe of suitable length produces four blades. Each blade is a circular segment with a center angle of 72°.

The runner design, with dimensions for a 1-foot-long runner, is shown in Figure 30; Figure 31 gives the nozzle design and dimensions. The dimensions can be altered proportionally for runners of other sizes. The shape of the nozzle can be made to suit penstock pipe conditions upstream from the nozzle discharge opening of 1¼ inches.

To calculate the principal turbine dimensions:

$$(B_1) = \text{width of nozzle (inches)} = \frac{210 \times \text{flow (cubic feet per second)}}{\text{runner outside diameter (inches)} \times \text{head (feet)}}$$

(B_2) = width of runner between discs = (B_1) + 1.0 inch rotational speed (revolutions per minute) =

$$\frac{862 \times \text{head (feet)}}{\text{runner outside diameter (inches)}}$$

The efficiency of the Michell turbine is 80 percent or greater, and therefore suitable for small power installations. Flow regulation and governor control of the flow can be effected by using a center-body nozzle regulator (a closing mechanism in the shape of a gate in the nozzle). A governor is expensive, but necessary for running an AC generator.

The application of Figures 28 and 29 is a typical example. For high heads the Michell turbine is connected to a penstock with a turbine inlet valve. This requires a different type of arrangement

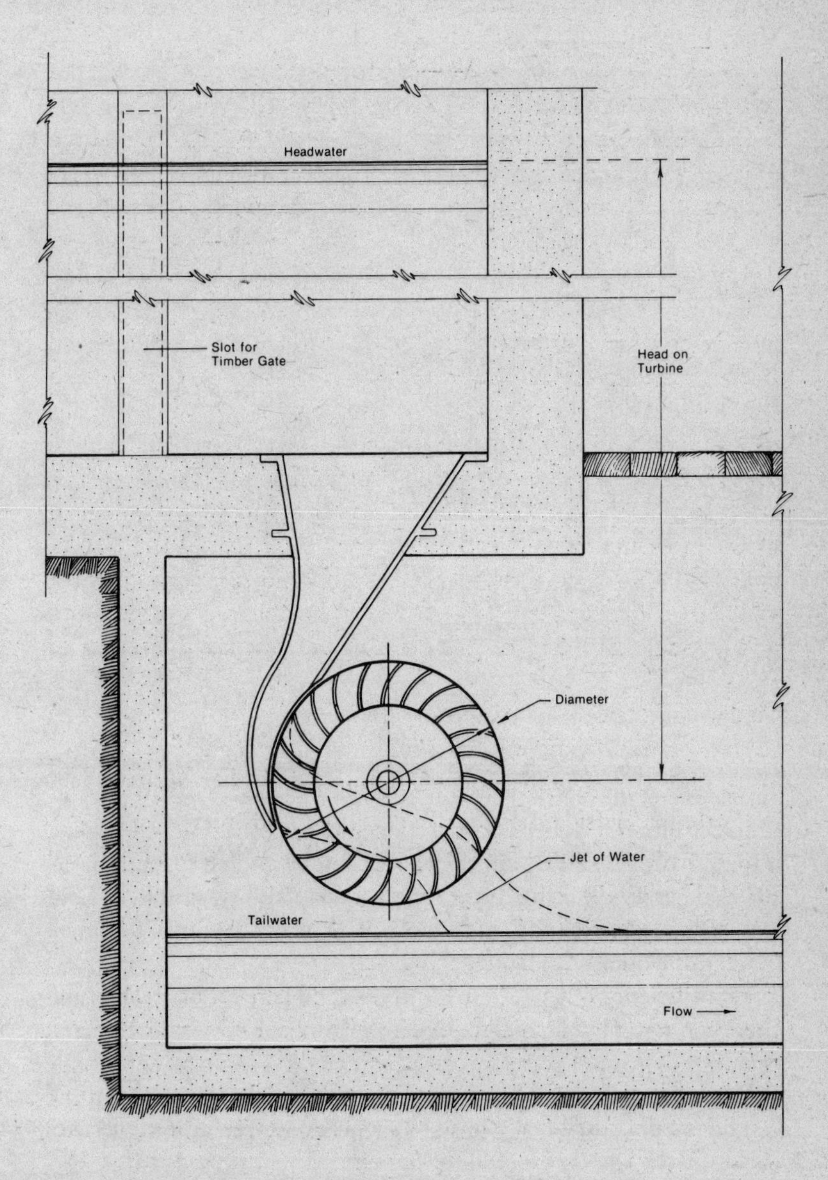

FIGURE 28. Arrangement for a Michell (Banki) turbine for low-head use without control (A).

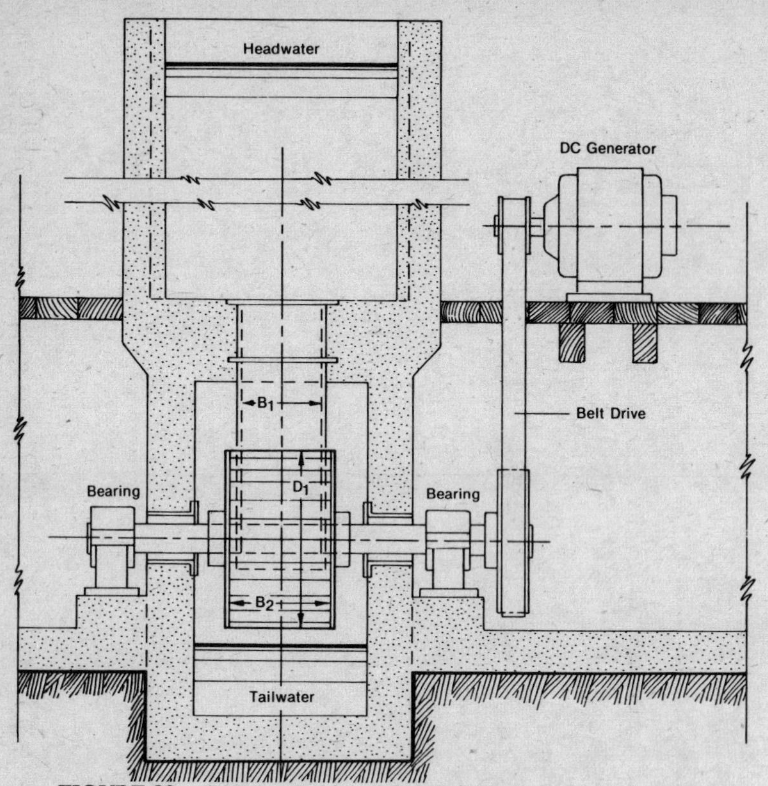

FIGURE 29. Arrangement for a Michell (Banki) turbine for low-head use without control (B).

from the one shown here. As mentioned before, the Michell turbine is unique because its B_1 and B_2 widths can be altered to suit power-site traits of flow rate and head. This, its adaptability, simplicity, and low cost, make it the most suitable of all water turbines for small power developments.

Centrifugal Pumps and Propeller-Type Pumps

The use of centrifugal pumps or propeller-type pumps as turbines should be explored before all other alternatives, provided that DC electricity can be used (See Figures 32 and 33.) Centrifugal pumps and propeller-type pumps are relatively low in cost and are available

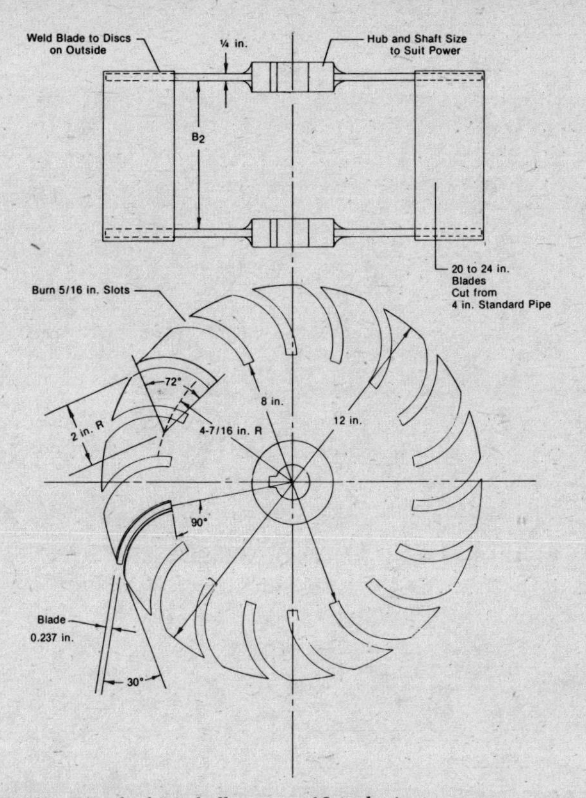

FIGURE 30. Detail of Michell runner, 12-inch size.

in many sizes. Manufacturers can quote the proper unit if head and flow are given.

These pumps can be used to produce AC electricity also, but with increased cost. In this case, a butterfly valve is used as the turbine-inlet valve; and the valve can be regulated by a small water-turbine governor.

The help of an engineer should be sought in modifying these pumps for use as turbines.

Water Wheels

Water wheels date back to biblical times but are far from obsolete. They have certain advantages which should not be overlooked. In

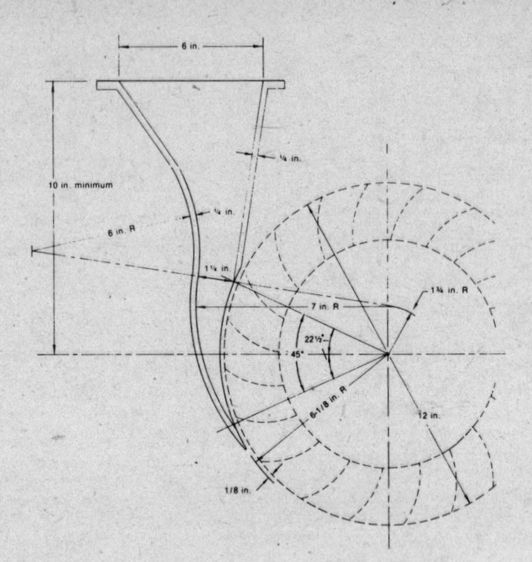

FIGURE 31. Detail of nozzle for 12-inch Michell runner.

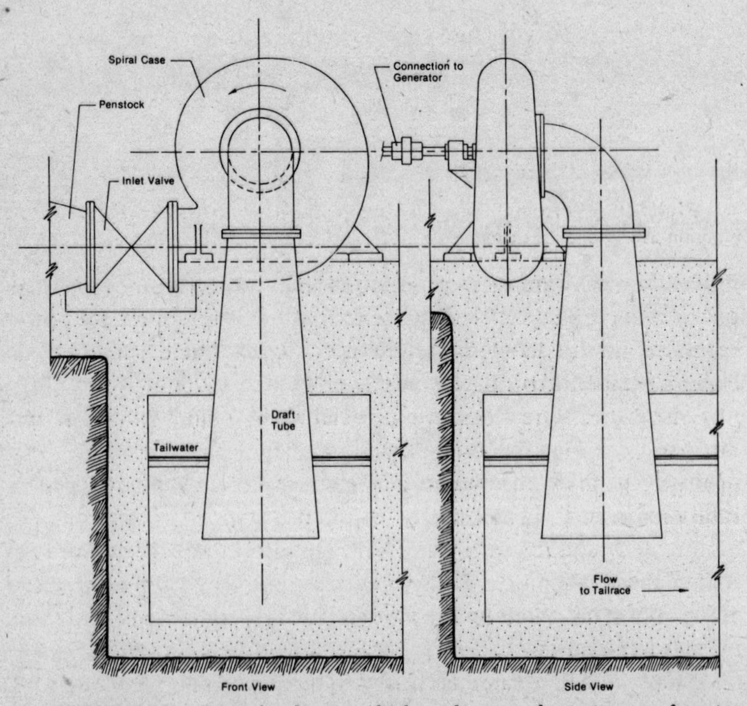

FIGURE 32. Centrifugal pump which, with reversed rotation, can function as a water turbine.

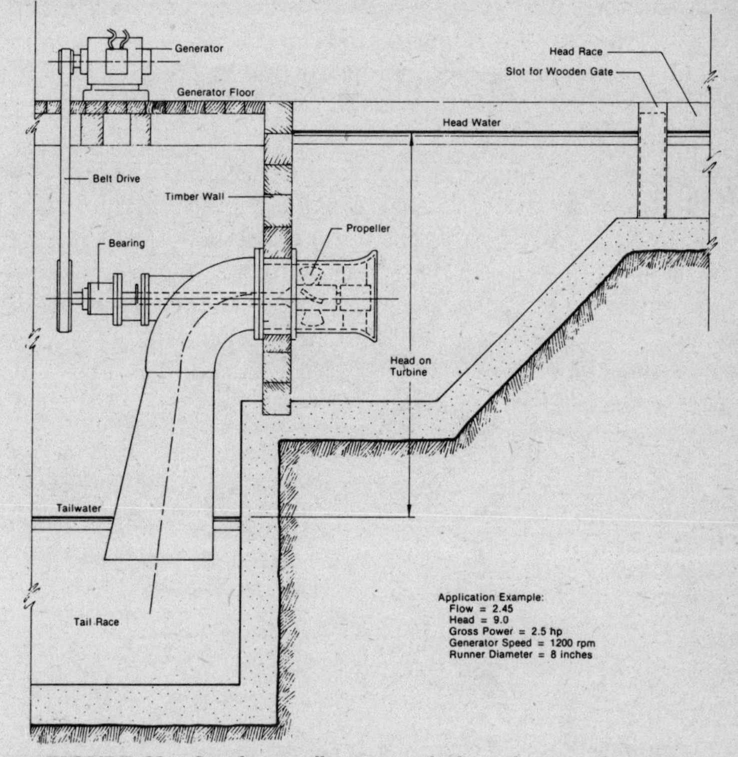

FIGURE 33. Simple propeller pump which, with reversed rotation, can function as a propeller turbine.

some cases, they are more economical than water turbines for small power requirements. It is possible to make a water wheel for power requirements up to 10 horsepower in places where there are no elaborate manufacturing facilities.

Water wheels are attractive, especially where fluctuations in flow rate are large, but speed regulation is not practical. They are used primarily to drive machinery which can take large fluctuations in rotational speed. They operate between 2 and 12 revolutions per minute and require gearing and belting (with inherent friction loss) to run most machines. Thus, they are most useful for slow-speed applications, e.g., flour mills, some agricultural equipment, and some pumping operations.

A water wheel, because of its rugged design, requires less care than a turbine does. It is self-cleaning, and, therefore, need not be protected from debris (leaves, grass, and stones). The two main types of water wheels are the overshot and the undershot.

OVERSHOT WATER WHEEL

The overshot water wheel may be used with heads of 10 to 30 feet, and flow rates from 1 to 30 cubic feet per second.

The water is guided to the wheel in a timber or metal flume at a water velocity of approximately 3 feet per second. A gate at the end of the flume controls the flow to the wheel and the jet velocity, which should be from 6 to 10 feet per second. To obtain this velocity, the head (H_1 in Figure 34) should be 1 to 2 feet.

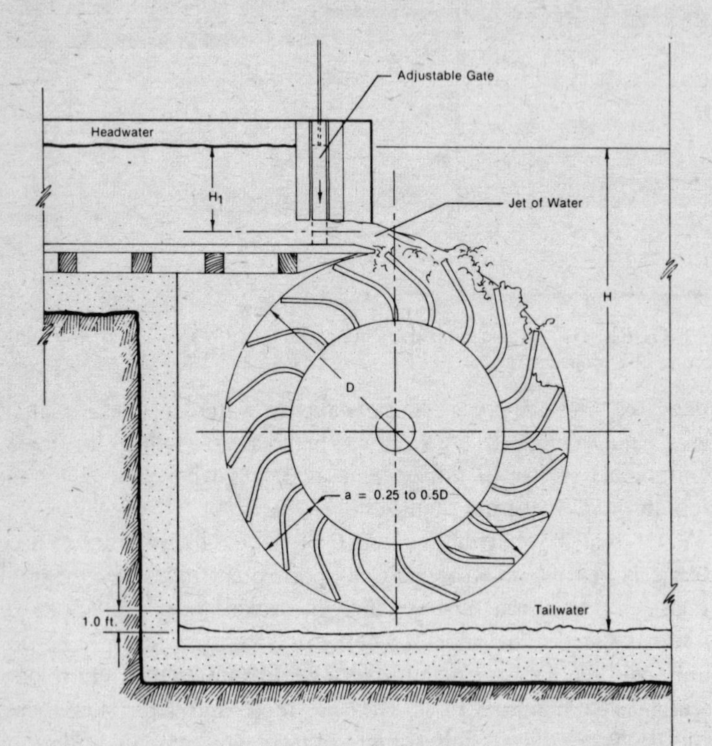

FIGURE 34. Overshot water wheel.

Wheel width depends on the amount of water to be used. The discharge will be 1 to 2 cubic feet per second for a flume width of 1 foot. Wheel width must exceed flume width by about 1 foot because of jet expansion.

The efficiency of a well-constructed overshot water wheel can be 60 to 80 percent.

Undershot Water Wheel

The undershot water wheel (Figure 35) should be used with heads of 1.5 to 10 feet and flow rates from 10 to 100 cubic feet per second. Wheel diameter should be three to four times the head. Rotational speed should be 2 to 12 revolutions per minute, with the higher speed applying to the smaller wheels. For each foot of wheel width, the flow rate should be between 3 and 10 cubic feet per second. The wheel dips from 1 to 3 feet into the water. Efficiency is in the range of 60 to 75 percent.

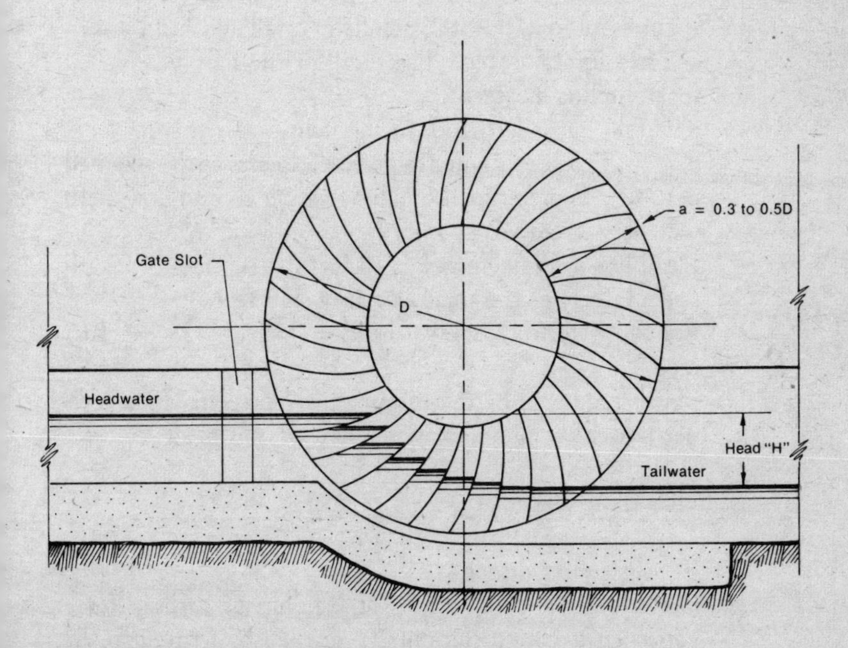

FIGURE 35. Undershot water wheel.

Examples

Mission Hospital

1. Requirements: 10 kilowatt light and power plant.
2. 10 kilowatts is 13⅓ horsepower.
3. The gross power required is then about 27 horsepower.
4. A stream in hilly territory can be dammed up and the water channeled through a ditch ½ mile long to the power plant site.
5. A penstock 250 feet long will take the water to the turbine.
6. The total difference in elevation is 140 feet.
7. Available minimum flow rate: 1.8 cubic feet per second.
8. The soil in which the ditch is to be dug permits a water velocity of 1.2 feet per second.
9. Table 7 gives $n = 0.030$.
10. Area of flow in the ditch = 1.8/1.2 = 1.5 square feet.
11. Bottom width = 1.5 feet.
12. Hydraulic radius = 0.31 × 1.5 = 0.46 feet.
13. Figure 11 shows that this results in a fall and head loss of 1.7 feet for 1,000 feet. The total for the half-mile (2,640 feet) ditch is 4.5 feet.
14. The fall that is left through the penstock is then: 140 − 4.5 = 135.5 feet. Figure 13 gives 5.7 inches as the required penstock diameter for 1.8 cubic feet per second flow at 10 feet per second velocity.
15. Head loss in the penstock is 10 feet for 100 feet of length and 25 feet for the total length of 250 feet.
16. For the water turbine: Net head = 135.5 − 25 = 110.5 feet.
17. Power produced by the turbine at 80 percent efficiency:
$$\text{net power} = \frac{\text{minimum water flow} \times \text{net head}}{8.8} \times \text{turbine efficiency} = \frac{1.8 \times 110.5}{8.8} \times .80 = 18 \text{ horsepower.}$$
18. Consult Table 8. The cost of a pump or turbine for a particular situation can only be learned by writing to the various manufacturers.

Availability of Manufactured Turbines

Small hydraulic turbines, and even more the governors for regulating these turbines, are difficult to obtain because the demand for these products has diminished to a considerable extent in the last 20 years. And manufactured water wheels are completely off the market. Of the remaining number of manufacturers of small turbines and governors only one exists in the United States, and two are known by the author to exist in Europe.

James Leffel & Company is located in Springfield, Ohio. Their booklet, Leffel Pamphlet A, "Hints on the Development of Small Water Power," is available on request. It is a very useful supplement to the information in this section. Its description of Leffel's small vertical Samson turbine is very complete. This turbine is available in sizes from 3 to 29 horsepower. The company maintains an engineering department which stands ready to assist in planning and design of the entire installation.

This company also manufactures a complete unit, called Hoppes Hydro-electric Unit, which is useful in isolated locations where power demand is small. It comes in sizes of 1 to 10 kilowatts. A Leffel bulletin describing this unit tells just what data you will need to submit in order to get the unit that best suits your needs.

The Michell (or Banki) turbine is manufactured exclusively by the Ossberger-Turbinenfabrik of Weissenburg, Bayern, Bavaria, Germany. This turbine is made in sizes ranging from 1 to 1000 horsepower. The company has an impressive record of installations, many in less-developed countries.

Ossberger-Turbinenfabrik is very responsive to requests for information. It furnishes, without charge, a considerable amount of data which is translated into English. The simple design of the Michell turbine makes it a favorite for remote regions and is priced lower than corresponding Francis and impulse-type turbines. Its governor, developed by Ossberger, is also very reasonably priced.

A third company which manufactures turbines and governors for turbines, but does not sell packaged units or the necessary electrical equipment, is the Officine Buehler, Taverne, Canton Ticino, Switzerland. They are in the small turbine field, and they manufacture all types except Michell. Their workmanship is of the highest quality, and their engineering is superb. Like the other companies, they assist prospective customers in planning their installations.

Hydraulic Rams

Volunteers in Technical Assistance (VITA)

How They Work

A hydraulic ram is a simple device, invented in the early 19th century. It uses the power from falling water to force a small portion of the water to a height greater than the source. Water can be forced about as far horizontally as you desire, but because of friction, greater distances require larger pipe. There is no external power needed, and the ram has only two working parts. The only maintenance needed is to keep leaves and trash cleaned away from the strainer on the intake and to replace the clack and non-return or delivery valve rubbers if they get worn. There is almost no expense except for the original cost. And a home-built ram costs about one-tenth the cost of a manufactured one.

Two things are needed to make the ram work: (a) enough water to run the ram and (b) enough height for water to fall through the drive pipe to work the ram. A small amount of water with plenty of fall will pump as much water as a greater amount of water with only a little fall. The greater the height to which the water must be raised, the less water will be pumped, under a given set of circumstances.

Water may come from a spring on a hillside or from a river. It must be led into a position from which it can pass through a relatively short supply pipe to the ram, at a fairly steep angle (about 30 degrees below the horizontal is good). Often a catch basin or cistern is used as the source for the drive pipe, but an open ditch could be used. (See Figure 36.) Be sure to put a strainer on the top of the drive pipe to keep trash out of the pipe and ram.

The ram works on simple principles. The water starts to run down through the drive pipe, going faster and faster until it forces the automatic valve or clack to close suddenly. The weight of the moving water suddenly stopped, creates very high pressure, and

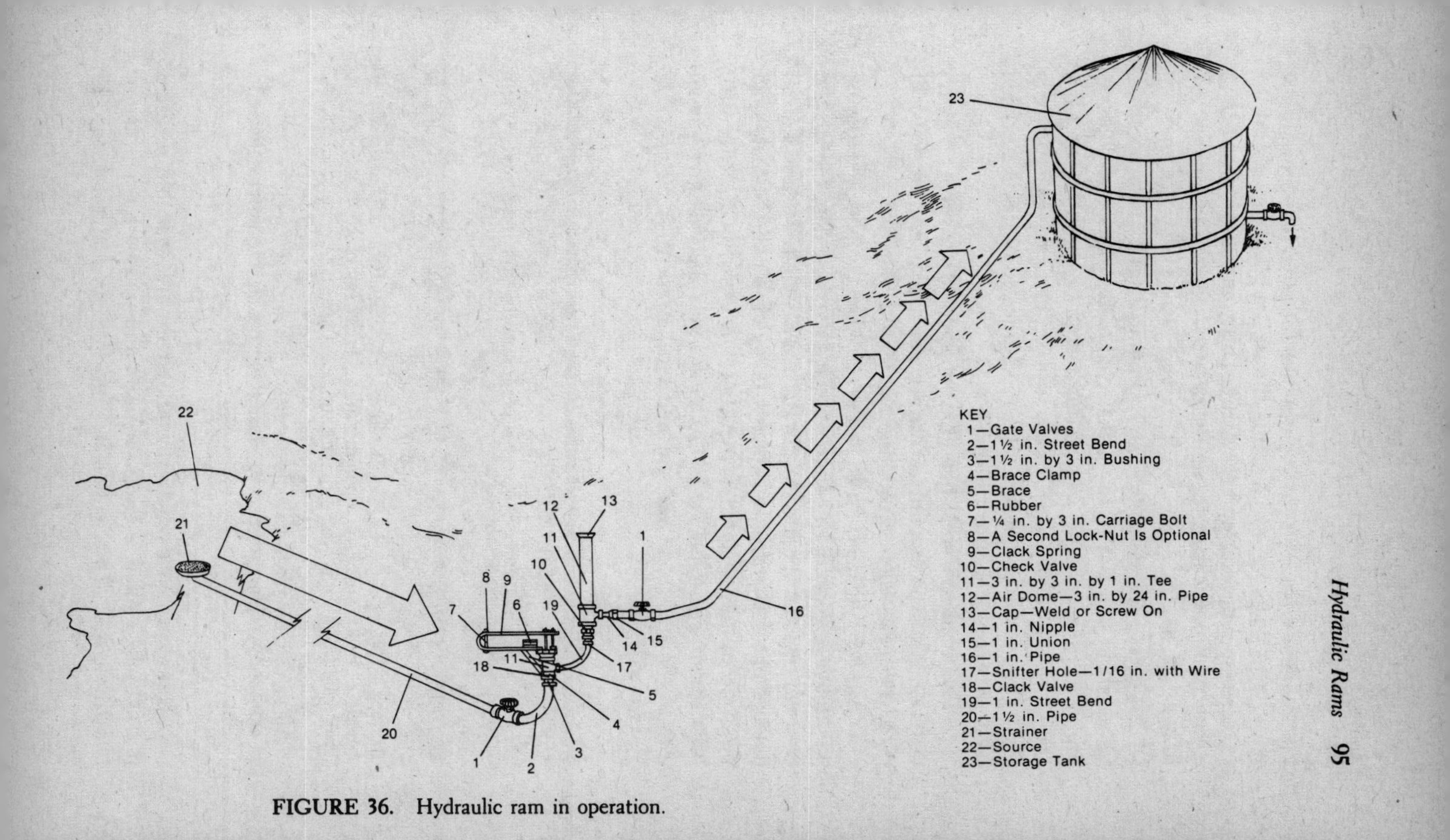

FIGURE 36. Hydraulic ram in operation.

forces some of the water past the non-return or delivery valve and into the air chamber, compressing the air more and more until the energy of the moving water is spent. This compressed air acts as a spring and forces the water up the delivery pipe to the storage tank in a steady stream.

It takes a lot of falling water to pump a little water up a hill. Often about one part in ten is delivered to the storage tank at the top of the delivery pipe. The snifter hole wastes a bit of water but takes in a bubble of air with each stroke. This is necessary to keep air in the air dome, which must not get plugged or it will get filled with water and the ram will stop.

The small ram works best at about 75 to 90 strokes per minute, depending on the amount of drive water available. The slower it goes, the more water it uses and the more it pumps.

Any working fall from 18 inches to 100 feet can be used to work a ram, but in general, the more working fall you obtain, the less the ram will cost and the less drive water it will require to raise a given amount of water. If there is plenty of water, a fall of 4 feet could be made to raise water 800 feet, but this would be an expensive installation. The following is a rough formula that will give you an idea of the amount of water which can be raised:

$$\frac{\text{Driving water per minute in gallons or liters} \times \text{twice the working fall in feet or meters}}{3 \times \text{vertical lift above ram in feet or meters}} = \text{Amount of water raised by the ram}$$

EXAMPLE: Working fall = 18 feet, lift above ram = 200 feet, driving water = 160 gallon/minute

$$\text{Water raised} = \frac{160 \times 2 \times 18}{3 \times 200} = \text{9.6 gallons per minute or 13,824 gallons per 24 hours}$$

This would require a No. 7 Blake ram.
100 gallons falling 10 feet would elevate 10 gallons to 80 feet.
100 gallons falling 5 feet would elevate 1 gallon to 300 feet.
Double the working fall and you just about double the water delivered.

Unless you have practically unlimited water available, measure it exactly by making a temporary dam and putting a large pipe or two through it. Then catch and measure the water for, say, 15

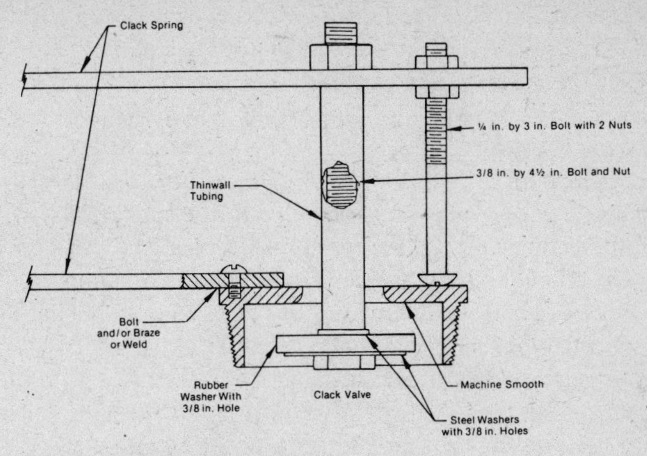

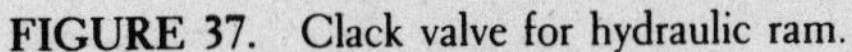

FIGURE 37. Clack valve for hydraulic ram.

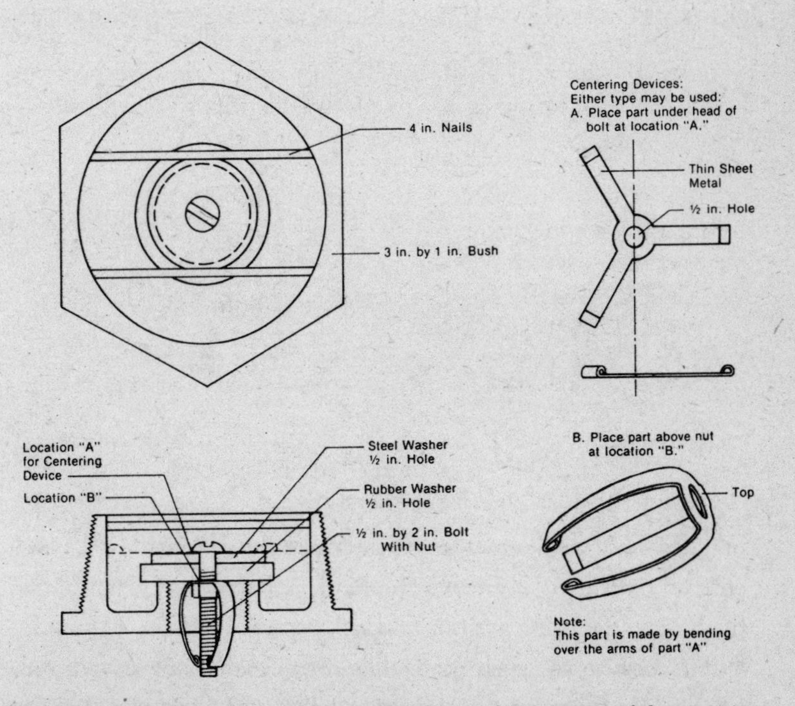

FIGURE 38. Check valve for hydraulic ram. (The purpose of the centering device—see detail on the right—is to prevent the moving assembly from slipping off-center or to the side.)

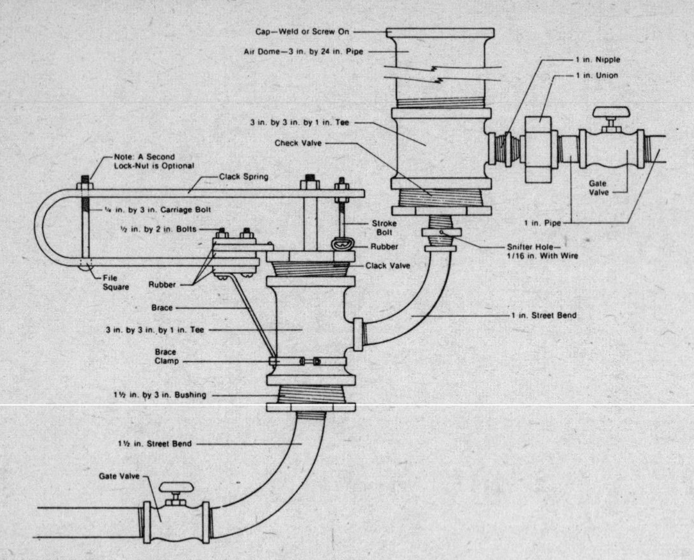

FIGURE 39. Complete assembly of hydraulic ram.

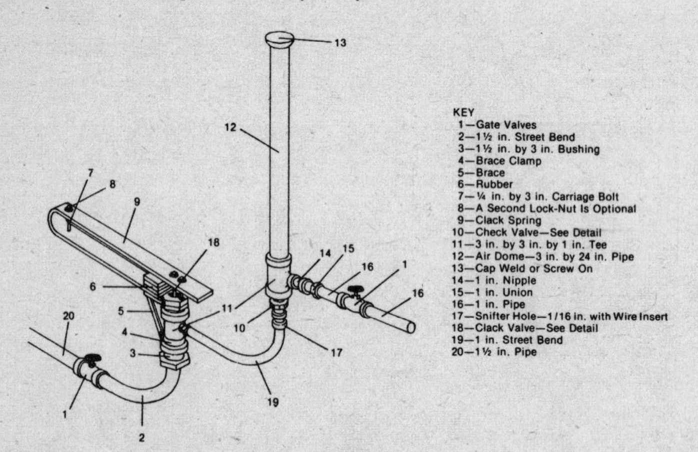

FIGURE 40. Assembly drawing of hydraulic ram.

minutes. Next, sight along a carpenter's level to the top of a 10-foot pole set on the ground down the hill at a lower level. Then move the level to the pole's position and sight again to the top of the pole, finding how many levels or fractions you have; this will give you, when added together, the amount of fall for the drive pipe. Do the same for the height to which the water must be raised. This height is measured from the ram level.

Building the Ram

Start by building the clack valve. (See Figure 37.) If you do not have a metal lathe, a machine shop will do the work for a small price. Chuck a 3-by-1-inch pipe bushing in the lathe and turn the inside smooth, where the clack strikes. Turn out the threads and eliminate any sharp edges. Drill two ¼-inch holes near the end of a piece of strap iron ¼-by-1½-by-3-inch, and using it as a template, drill and tap holes in the top of the pipe bushing. Grind off the galvanizing, then bolt the clack spring support solidly to the bushing and braze it.

Bend a 36-inch iron strap, 1½-by-⅛ inches, around a 2-inch pipe to make the clack spring. Drill two ½-inch holes through the end and also through both the support and the two short pieces to make up the pad as shown in Figure 37.

Cut pieces of rubber inner tube and assemble the sandwich. This sandwich will keep vibrations from breaking the support off the pipe bushing. A brace can be added for additional support, but it is not absolutely necessary.

The clack valve itself is made up of a rubber disc and metal washer ⅜ inch smaller than the inside of your bushing and assembled on a ⅜-by-4½-inch bolt. One of the best sources of rubber for the disc is an old tractor tire—it shows no wear at all after eight months' use. Cut it on a band saw and sand it flat and even on a disc sander with coarse paper. A similar piece of rubber is used for the check valve.

Slip a washer over the bolt and a short length of thin wall steel tube (¾ inch o.d. conduit) with the ends filed exactly square. Then slip it through a hole in the clack spring. Adjust it by bending, so that the rubber clack strikes true and doesn't rub on the sides of the bushing.

Drill a hole for a carriage bolt to adjust the stroke of the spring, then drill a pair of holes about 3 inches from the round end of the spring for a tension bolt. If the bottom hole is filed square to fit the underside of the bolt, it will not turn when adjustments are made.

The check valve (Figure 38) is similar in construction, but a

½-by-2-inch galvanized bolt is used. Machine the lip true where the valve rests, but do not cut it down farther than necessary. This gives a bit of clearance for the water to pass. Drill two holes on each side of the middle for a 4-inch common nail to pass just above the valve metal washer, to keep it in place. Leave enough clearance so the valve can open just about $^1/_{16}$ inch. Spread the bolt with a center punch just below the nut, so that the nut can't work loose. Cut the nails off and file threads across their ends so the bushing will screw into the tee above it.

Just one other small job is necessary before assembly: Drill a $^1/_{16}$-inch hole in the center of the 1-inch nipple just below the check valve. Then bend a piece of copper wire to the shape of a cotter pin and insert it from the inside of the nipple with long nosed pliers. Spread the outside ends. This copper wire restricts the jet of water coming out, yet moves enough to keep the hole clean.

The air dome can be a 2-foot length of 3-inch pipe, threaded on both ends with a cap or a welded plate on the top end. The dome must be airtight at great pressure. Coat the inside of the pipe with asphalt paint to protect it from rust and to seal any small leaks in the weld. Let it dry in the sun while assembling the rest of the ram.

ASSEMBLY

Use plenty of good grade pipe joint compound, both on inside and outside threads. Screw components together firmly, but not excessively tight, and leave them in the correct position for your installation. (See Figures 39 and 40.) Set the ram reasonably level. The snifter hole must be immediately below the air dome so that the bubbles will go up into the dome. Clack and check valves must be free from binding and touch evenly all around. The old tractor tire rubber with some fabric on the back seems to be just the right toughness and resiliency to last a long time—much longer than either gasket rubber or live rubber.

There is no reason at all to mount the ram in concrete as has been suggested by some. In fact, it is very convenient to have a "portable" ram and to be able to shut off the two valves, loosen the unions, and take the ram to the shop for cleaning and painting. Painting, of course, doesn't improve operation, but it does improve your ram's appearance.

A bit of rubber stretched over the head of the stroke bolt will help to quiet the ram. Adjust the spring tension bolt and stroke bolt together to get the best period for your particular ram. Support the drive and delivery pipes so they don't bounce and vibrate.

The ram described here is a small one, but larger rams can be built. VITA has built two with 3-inch drive pipes and correspondingly larger ram parts. One of our larger hydraulic rams lifts water about 150 feet and drives it through 3600 feet of pipe.

Installation and Adjustments

The drive pipe should have a strainer on the top made of ½-inch coffee tray wire, hardware cloth, or anything similar. This wire will keep out trash, frogs, and leaves, all capable of clogging up the ram. The drive pipe should be 1½ inch or larger (we use 2-inch pipe) and, if possible, it should be new, solidly put together, straight, and well supported throughout its length. A gate valve on the drive pipe about 4 feet from the ram is a great convenience, but is not necessary. Another gate valve on the delivery pipe is almost a necessity since it will prevent the entire delivery pipe from draining whenever the ram is cleaned. The ram should be connected to the delivery and drive pipes by unions so it can be removed for cleaning. (See Figure 36.)

If it is desirable to use two rams, they must have separate drive pipes, but the delivery pipes can be joined, provided the pipe is large enough to carry the increased quantity of water.

The delivery pipe should lead off from the ram with about two lengths of 1-inch galvanized iron pipe. From there ¾-inch plastic pipe can make up the remainder of the delivery pipe. The iron pipe will give the ram better support, but plastic pipe is smoother inside and can be a size smaller than the iron pipe. Also, plastic pipe is cheaper, but it must be protected from mechanical injury and sunlight.

The length of the supply line must be at least three times the length of working fall. If it is shorter the ram will stop when the tap is turned on. (A float valve, however, might prevent this from happening.)

The small bolt at the end of the clack spring controls the length of the stroke of the ram. The bolt at the back (rounded) end of the spring controls the tension of the clack spring. (See Figure 38.) Experiment for the best length of stroke and tension for your set of conditions. Adjust the length of stroke first, then the spring tension. The greater the tension and length of stroke, the slower the ram will work and the more water it will pump, but it will take more water to keep it working.

For a reliable, constant supply of water, lead all the water from the ram directly into a storage tank and use it from there. The overflow can be used to irrigate your garden or field.

IF ACTION IS FAULTY:

• See that the clack valve closes squarely, evenly, and completely. If it does not, the clack spring may have been bent somehow, and it will have to be straightened.

• See that the clack valve does not rub on the front, side, or back of the valve body.

• Check for trash in the ram, delivery valve, or in snifter hole.

• Check to see that the air dome is not filled with water. If it is filled with water, the ram will knock loudly and one or more parts may break. The snifter should allow a small amount of air to enter between each of the strokes to keep the dome full of compressed air.

• Check the rubber clack and delivery valve for wear or looseness.

• If drive water is in supply, speed up the stroke by loosening spring tension and shorten the stroke by lowering the stroke-adjusting bolt. More water is delivered by a faster stroke and continuous running than a slower stroke that stops every day.

• Check for leaks in the drive pipe. If air bubbles come out of the drive pipe after it has been stopped for a while, air is leaking into the drive pipe and is interfering with the ram action.

3 WOOD POWER

Heating and Cooking with Wood

Ken Kern

The now-popular phrase, *do your own thing,* coined more than a century ago by Ralph Waldo Emerson, aptly expressed traditional Yankee values. More and more people today, seeking an alternative to "contemporary living," have romantically turned their gaze back to those "fine times" of realism and self-reliance. But with all its earthy nostalgia, to do your own thing in 19th century Emersonian fashion would be extremely arduous and, assuming 20th century men and women could and did succeed in mastering such a lifestyle, it would still leave them esthetically, culturally, and socially underfed. Consequently, at some point in our development during the last hundred years, Emerson's philosophy was scrapped in favor of a more comfortable and *technically* challenging existence: it was found that machines could do our thing for us.

In 19th century Massachusetts, Emerson did his own thing by chopping fireplace wood to keep his house warm. Fuel gathering was slow and difficult, fireplaces totally inefficient, houses inadequately insulated and improperly orientated. Little wonder his neighbors went forward to evolve new and improved methods of heating their homes. Typical of 20th century machine technology, heating methods were developed to give automatically controlled, equal, and evenly distributed heat flow by way of some umbilical fuel line. Dependence on the factory-produced furnace and radiator or ductwork, and on the distant fuel supply would, of course, be abhorrent to Emerson: not only his self-reliance reduced to a thermostat setting, but one's very freedom and independence in the hands of centralized fuel monopoly and/or government control.

A modern technologist, eager to succeed, would find little reward in re-designing and properly engineering Emerson's wood-burning system to make it a viable house-heating solution. Rather, the modern challenge lies in thermocouples and microswitches, heat pumps and convector converters. Even devising a sophisticated solar collector-storage system or fancy generator to produce methane fuel can be more fun and present a greater challenge to an engineer's superior mentality.

Actually there has been worldwide *behind-the-scenes* technological development to make wood-burning a viable method of house heating in colder climates. I stress behind-the-scenes because wood-burning technology in this country has been very much out of the heating and air-conditioning mainstream. Much of our knowledge on the subject has come out of or has been directed toward emerging third-world nations.

And the knowledge that we have gained in the past few decades would warm Emerson heartily: he could have created a comfortable living environment, heated entirely by prunings and thinnings from his woodlot, by merely combining the skills and materials then at his disposal with our present technical knowledge.

Emerson's New Concord House

Accordingly, I have re-designed Emerson's Concord house and heating system so that the entire unit can be fabricated by anyone well motivated toward mastering his existence. The revised h design is mine, but I have borrowed ideas for the building an equipment design developed through the years by others: Frank Lloyd Wright offered ideas on earth-berming insulation (using packed earth as insulation against the cold north wall); Wendell Thomas contributed greatly with his "no-draft floor" inventions; from India, Dr. Billig assisted with his plunger pile floor ideas; Peter van Dresser recalled the need and value of suntempering; the English scientist, Dr. P. O. Rosin contributed a wealth of information on the aerodynamics of open fires; Jack Bays showed how to make some truly low-cost, versatile building materials from waste materials; Scott Nearing demonstated how anyone with minimal skill can build well-insulated, native-stone walls. These are only a few of the people concerned with new-era housing; their expertise is described in detail in my book, *The Owner-Built Home.*[1]

At this point you may begin to realize that to efficiently heat Emerson's new Concord house with wood, much concern must be given to walls, floor, roof, insulation, and solar exposure, as well as to the wood-burning fixtures themselves. Glance for a moment at the plan: Notice, first of all, the circular shape. The northern half is curved so as to better withstand hydrostatic wall pressure exerted as a result of earth-berm insulation: the southern half is curve-angled and sloped so as to achieve maximum solar-ray incidence—from southeast to southwest radiants. The circular form also insures optimum warm air circulation, emanating from the centrally located heat source. The circulation system itself employs registers along the perimeter of the floors where the floor joins the outside walls. Heated air, cooling as it rises to the higher levels, falls and flows downward along the outside walls, through the registers and thence *under* the floor and back to the heating unit for recycling (as indicated in the drawing by flow arrows).

The No-Draft Floor

Additionally, this system provides for the unheard of "no-draft floor." Sudden rushes of cold air from open doors fall into the ventilation inlets (see Figure 42) by virtue of their weight, instead of blowing across the lower area of a room as they do in houses of

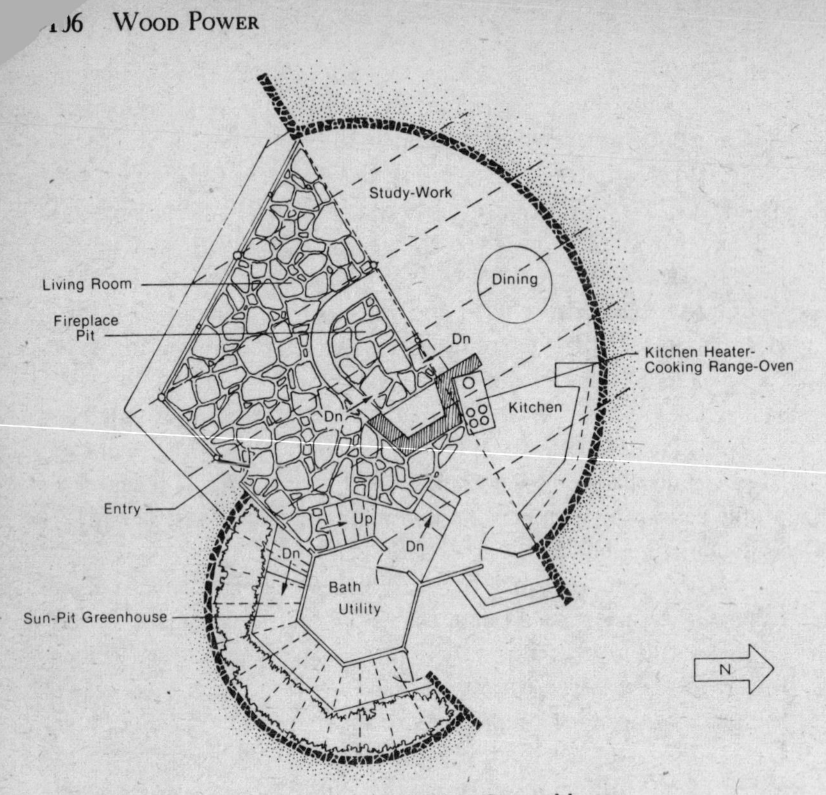

FIGURE 41. Floor plan for Emerson's New Concord house.

contemporary "design." Air space below the floor is achieved by casting the floor on loose, uncompacted earth fill between concrete plunger piles spaced every 3 feet each direction. The piles are made with a crowbar driven into the ground to a depth of 3 feet, then filled with concrete. After a few weeks the loose earth settles and an air space is formed under the slab which finally rests on the concrete piles. (See Figure 43.) The heat circulating through the stove or the inner chamber of the fireplace, across the ceiling, down the walls, and thence under the floor and back to the heat source is *convected* heat.

The Sunken Hearth

An equally valuable heat source is obtained from direct *radiation* around the fireplace hearth or wood heater. A 12-inch sunken

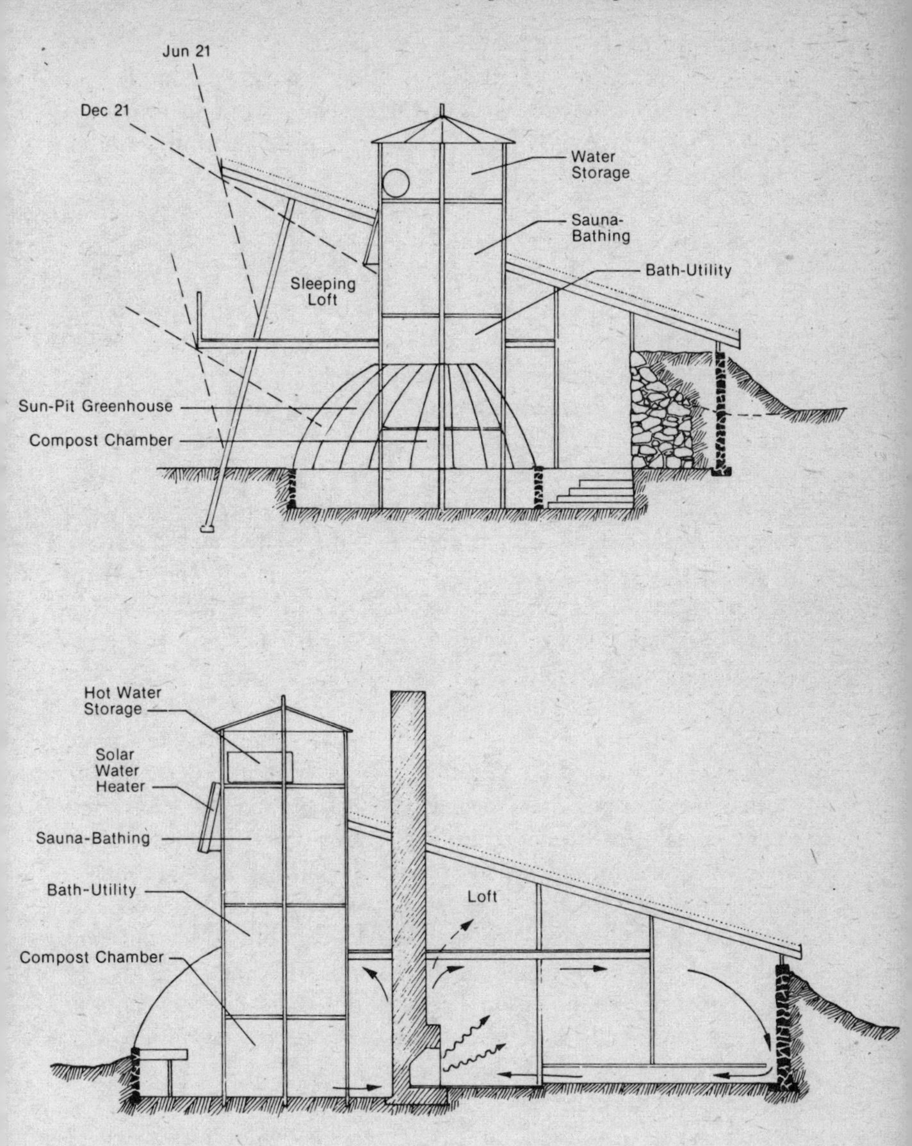

FIGURE 42. Side view of Emerson's New Concord house. Above, N-S section; below, E-W section.

fireplace hearth puts the heat radiation at floor level where it is most needed. It also creates an intimate seating arrangement and lessens the danger of flying sparks since the trajectory of a spark from a sunken hearth is considerably shorter than that from a raised one.

Ventilation and Insulation

A smooth and optimum warm-air movement is possible in this house only if exterior drafts can be controlled and adequate insulation provided. Windows accordingly are fixed; ventilation is provided by separate louvre openings. South-facing windows slope to an angle commensurate with the latitude of the Concord site. Thus, during the winter equinox, solar rays strike the glass at a near right-angle incidence, providing maximum penetration to the dark, slate-covered masonry floor. As the floor becomes heated by the sun, air below the floor expands and creates an air movement there. With all warm-air outlets located at the fireplace, a down-draft principle draws cool and moist air from the exterior window and wall area, then under the floor and up at the fireplace outlets.

After sunset, or during winter storms, insulated panels—old-fashioned "shutters" if you will—hinge over the glass windows. Insulation of the cold north wall is provided by earth-berming the masonry wall. Elsewhere—on walls, ceilings, and roof—a heavy coating of asphalt-clay-fiber mixture is plastered onto wire mesh. (This mixture consists of equal parts of red clay and ground [through a hammermill or shredder] fiber, such as sawdust, straw, cardboard, corn cob, rice hull, etc. To this is added enough bitumul [emulsified asphalt] and water to stabilize the compound.) The same ground fiber is packed as insulation between the wall and between the roof-framing members. An 8-inch layer of sod covers the exposed roof surface.

Note on the plan that the sleeping facilities are provided in the loft. This is a sensible arrangement, heat-wise, especially in northeastern climate zones. A certain amount of heat naturally filters to the loft (registers are installed to control the amount). At bedtime the loft area is always comfortably warm.

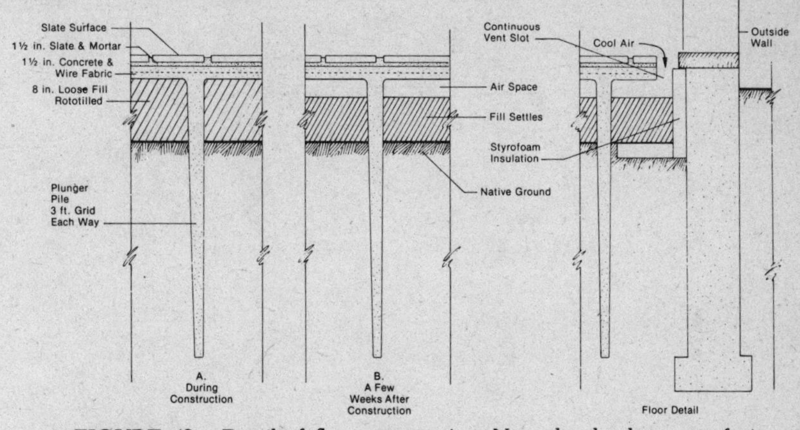

FIGURE 43. Detail of floor construction. Note the development of air spaces a few weeks after the floor is built.

The Fireplace

The nucleus of the Concord house heating system is its central fireplace. The key factor relative to optimum functioning of the fireplace is its circulating heat chamber. Sheet iron is 12 times as conductive as stone masonry. A metal heat chamber will emit quantities of conductive heat that would otherwise be lost through absorption in the surrounding masonry or lost to the atmosphere via the chimney. Patented metal fireplace liners (Majestic, Heatilator, Heatform) became available long after Emerson's time. All, except for Heatform, are poorly built out of thin, 14-gauge metal and, unfortunately, the liner is unreasonably expensive. In keeping with the do-your-own-thing theme, I recommend that people build their own fireplace liner from my plans. (See Figure 44.)

The metal chamber is cut, bent, and welded out of a single 3-by-9-foot piece of sheet iron. A damper is welded to a pair of hinges which in turn are welded to the smoke shelf. Cool air inlets are provided along the upper front and rear of the fireplace.

Air-intake control is the key to efficient fireplace combustion. The ignition of a correctly proportioned gas-air mixture will give complete combustion of wood and emit gases containing only the

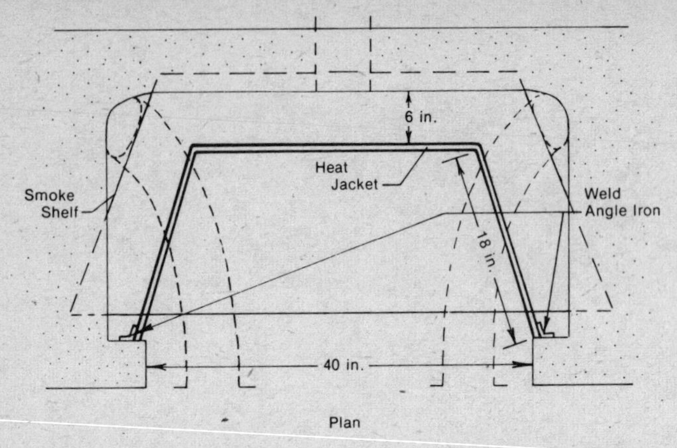

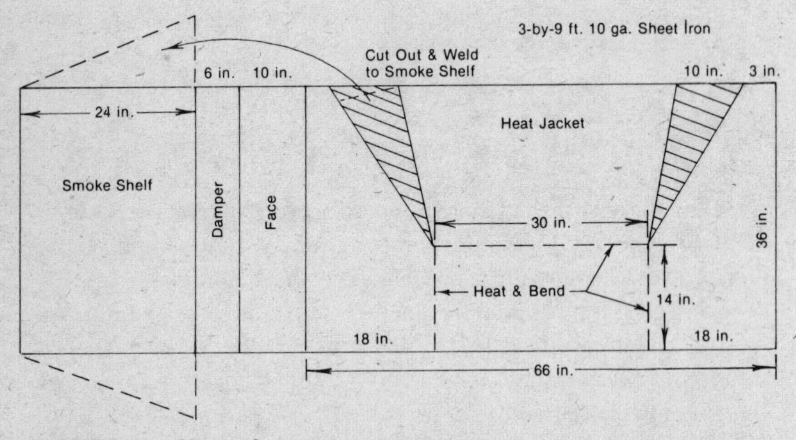

FIGURE 44. Heat jacket pattern.

non-combustible carbon dioxide, water vapor, oxygen, and nitrogen. When smoke and soot are observed coming out of a chimney, you can be sure that combustion is incomplete. What you actually see are small quantities of hydrocarbons and free carbon (soot) that are not burned. Thus, much of the heating capacity of the woodfuel is lost. The heat loss is twofold: in the hot gases that rise up the flue, and in the unburned combustible particles that escape with them.

It's common knowledge that green wood has less heating efficiency than well-seasoned wood. This is because freshly felled wood is 50 percent water, and a high moisture content interferes with the burning process. Water evaporating from wet wood forms a sheath of vapor around the fuel and blocks the entry of oxygen, thereby lowering ignition and combustion rates. The point of combustion of wet wood can be lowered to a reasonable level only by increasing the air intake. This, in turn, creates an excess draft through the fire with consequent heat loss via the flue. If you must burn wet wood you will need plenty of draft, but properly seasoned wood requires only a small amount of air for combustion. When burning dry wood, the draft must be controlled. This will now be explained.

A fire in a fireplace involves: (1) the motion of air toward the fire; (2) its passage through and over the fuel bed; (3) its admixture with combustion products; and finally (4) the flow of the mixture up the flue. Let's take these steps one at a time:

(1) The required air-volume flow toward the fire for the average fireplace opening is about 3,000 cubic feet per hour—which amounts to about four complete air changes per hour in an average-sized living room. Now the amount of fresh air required for proper ventilating by a family of four has been established at about 1400 cubic feet per hour. Thus, a standard fireplace will cause the displacement of over twice the amount of room air required for optimum ventilation. Hence half the amount of air needed for the fire should be drawn through duct-work directly from outside the room and not be permitted to pass through the room to the fireplace.

This is especially important when fireplaces are installed in our modern, tightly constructed, efficiently weather-stripped houses, because the amount of air available for supplying the chimney draft is insufficient! A partial vacuum results which, besides starving the fire, tends to pull smoke and combustion gases back into the room. This vacuum also causes foot-chilling drafts.

A sub-floor inlet for new air minimizes the cold air currents within the room. Combustion efficiency is thereby increased, especially if the incoming new air is heated before it reaches the firebed. This is best accomplished by routing the new air supply duct under the firebed with exit at the fire-grate. (See Figure 45 for illustration of this air flow.)

(2) A properly designed grate is critical to the second consideration—the passage of air through and over the fuel bed. A grate

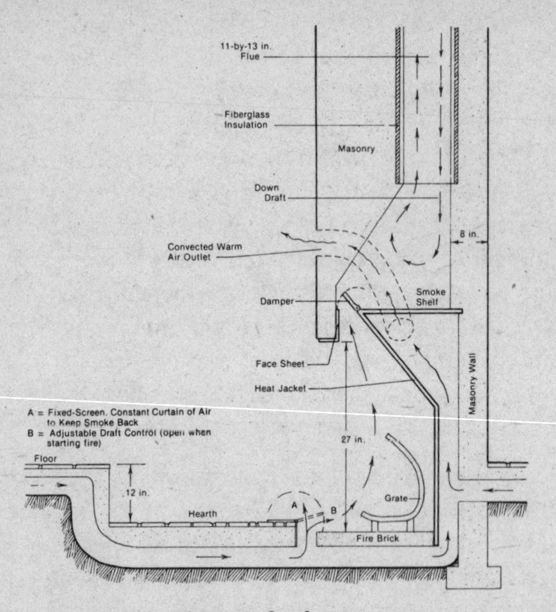

FIGURE 45. Cross-section of Kern's fireplace.

is used to raise the firebed a few inches above the hearth. It should be relatively small, as it is important that the wood charcoal which forms during combustion cover the grate completely. If the grate is too big so that too much wood is burned, the temperature of the gases escaping through the flue will be too high, resulting in heavy heat losses. On the other hand, too small a grate causes draft trouble and consequent incomplete combustion with the added risk of tar deposits in the chimney.

A 14-inch deep fuel grate adequately holds an ideally stacked triangle of three logs. It also insures that hot coals do not come in contact with the rear of the metal liner; thus, preventing red-hot metal-deteriorating temperatures.

One can easily fashion a fuel grate from a series of short pieces of 1-inch black iron pipe. (See Figure 46.) Horizontal members are attached to bent vertical pieces with a standard elbow connection. The six or more sections are then welded to two pieces of bar steel which have 3-inch high legs bent from both ends.

Fuel grates of similar design are made in many welding shops throughout the country, as well as by several manufacturers. One company claims that more than double the amount of heat output is achieved by using this "C" shape convection grate. Hot coals on the base and flames in contact with upper portions of the pipes cause heated air inside of each pipe to expand and to flow up and out into the room.

(3) We come, thirdly, to the admixture of ventilating currents with combustion products. The inside fireplace proportions are very important here. Deep fuel beds (long from front to back) produce more smoke than shallow beds, since there is scant combustion air at the back of the grate. Experience proves that the rate of smoke emission increases proportionally with the depth of the firebox, expecially in the early stages of firing. It is also apparent that radiation is more effective with a shallow firebox; with a deep box radiation is mostly upward. An inclined fireback, sloping forward as it goes up, also contributes to better radiation.

The throat opening should be sufficiently small—not over 4 inches wide—to constrain the effluent to pass at a speed high enough to discourage down-drafts. If the chimney throat is too large, then some cool air that does not participate in ignition or combustion will be drawn over the fire. This naturally results in increased smoke emission.

(4) The final aspect of a fireplace, the flow of gases up the flue, involves the chimney itself. A chimney performs a dual function: production of drafts and elimination of combustion residues. The chimney should be designed so that the draft is adequate to deliver just enough air for complete combustion within the firebox. Flue diameter or width should be as small as possible relative to its height. Its height is dependent upon the height of the building plus at least 3 feet above the roof, plus any additional footage required to clear the top of the flue of any obstructions within an approximate radius of 10 feet. (Figure 47 shows basic relationships between room size, fireplace proportion, and flue size for various chimney heights.)

The flue must also be insulated to maintain a high temperature and a near consistent temperature top to bottom. The masonry of which the chimney is constructed will not serve this purpose, as stone is a poor insulator. An insulation material like fiberglass should be used between the masonry and the flue lining. This improves draft and also retards condensation of tar and creosote

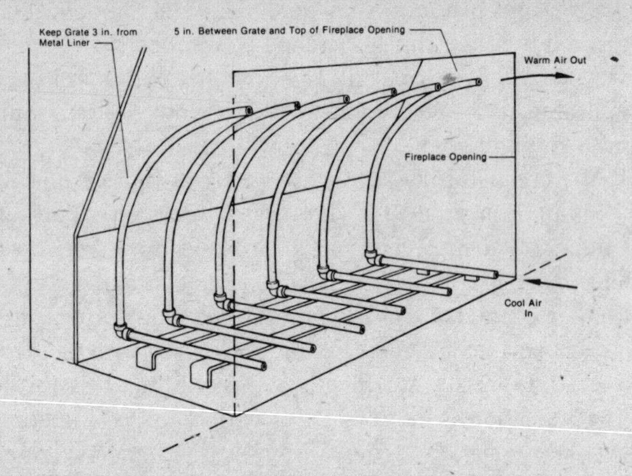

FIGURE 46. Fuel grate constructed from short pieces of iron pipe.

on the walls of the flue because condensation only occurs where low temperatures permit. Flue outlets for the wood range and auxiliary wood heater, as well as the fireplace, are also located in the masonry chimney, thus making it possible to use the safe and permanently installed fire-clay tile flue lining.

Some of the heat, which would otherwise be wasted up the flue, is used to heat water running through two sets of coiled pipes. One set of these pipes is attached to the inclined fireback of the fireplace and the other installed inside the heater stove. After being heated, the water is circulated to an insulated metal tank located within the masonry core.

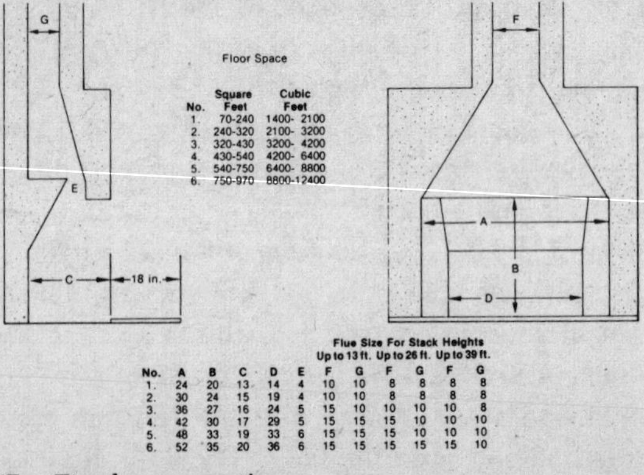

Floor Space

No.	Square Feet	Cubic Feet
1.	70-240	1400- 2100
2.	240-320	2100- 3200
3.	320-430	3200- 4200
4.	430-540	4200- 6400
5.	540-750	6400- 8800
6.	750-970	8800-12400

						Flue Size For Stack Heights					
						Up to 13 ft.		Up to 26 ft.		Up to 39 ft.	
No.	A	B	C	D	E	F	G	F	G	F	G
1.	24	20	13	14	4	10	10	8	8	8	8
2.	30	24	15	19	4	10	10	8	8	8	8
3.	36	27	16	24	5	15	10	10	10	10	8
4.	42	30	17	29	5	15	15	15	10	10	10
5.	48	33	19	33	6	15	15	15	10	10	10
6.	52	35	20	36	6	15	15	15	15	15	10

FIGURE 47. Fireplace proportions.

The Development of Wood Burners

Woodburning-fireplace and heater-cooking stove development have a history not unlike that of the development of many other of man's material acquisitions, such as his dwellings and his transport vehicles. Primitive (Paleolithic) solutions for fireplace or stove were crude, but they worked. "Civilized" (Neolithic) models were gross, wasteful, and worked poorly. Faint glimmerings of a new era (Biotechnic) lifestyle shine through our present-day morass. The light is carried by a few individuals who are able to see beyond the modern confines to a better future ahead.

Count Rumford was certainly one of these early, biotechnic torchbearers. In 1800, he published his comprehensive essay, "Chimney Fireplaces." His main contribution was in the alleviation of the smoking fireplace. One fault to the fireplaces of his time, he correctly asserted, was due to too large a chimney throat. He argued that if the chimney throat is too large then some cool air, not effecting ignition or combustion, will be drawn over the fire. This naturally results in an increase of smoke emission. The throat opening should be sufficient to constrain, venturi fashion, the passage of the escaping gases and unburned particles at a speed high enough to discourage downdrafts. Traditional chimney throat standards call for an 8-inch opening. Rumford recommended 4 inches. (It should be noted, however, that there is an alternative to reducing the throat size. A damper control, installed in the flue, can curb undesirable downdraft eddies.)

Rumford introduced the inclined fireback which increases fireplace efficiency by providing a greater radiation area. For the purpose of breaking up the current of smoke in the event of chimney downdraft, the back smoke shelf of Rumford's improved fireplace ended abruptly—a practice strictly adhered to by fireplace masons to this day. (See Figure 49.) He was also first to give extensive study to inside-fireplace proportions. According to Rumford, the fire-space proportion requires the *back* of the recess to be equal in size to the *depth* of the recess. Deep fuel beds produce more smoke than shallow beds since there is scant combustion air at the back of the grate. Experience proves that the rate of smoke emission increases proportionally with the increase in depth of the firebox, especially in the early stages of firing.

As I have already indicated, it is apparent that radiation is more effective with a shallow grate. With a deep grate radiation is mostly upward. Higher temperatures and consequent decrease of smoke emission results when the fireback becomes inclined toward the fire —another practice strictly adhered to by contemporary fireplace builders.

Rumford had an uncanny comprehension of fireplace aerodynamics, and, although he successfully cured the smoking fireplace, his prototype was far from being the heat efficient unit it could be. For this reason I am dismayed to see Rumford's ideas presented, as in *The Forgotten Art of Building a Good Fireplace* (by Vrest Orton, Yankee Inc., 1969), as the final authority on fireplace design. This book presents an excellent historical treatment, but modern fireplace builders, following the diagrams of this early proponent of fireplace design, realize only the disservice of half-truths for their labors. Much work preceded Rumford's revelations, and many advances have followed the Count's efforts.

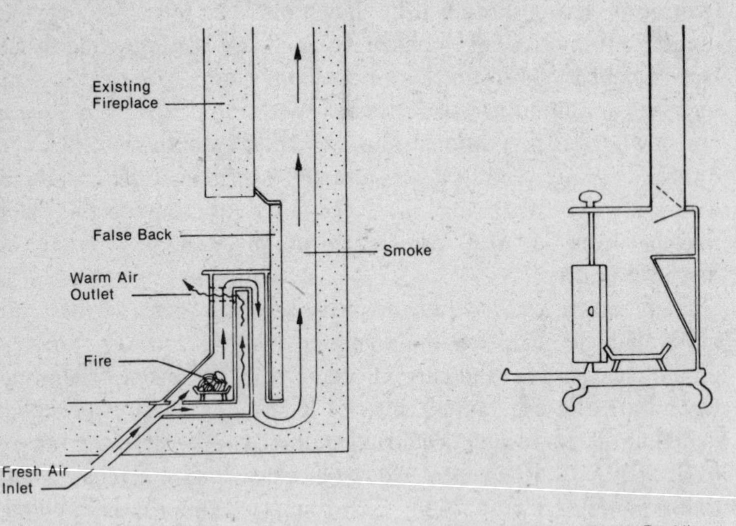

FIGURE 48. Two Franklin versions which can be installed in existing masonry fireplace.

For instance, Rumford apparently knew nothing about Louis Savat's heat circulating fireplace, installed in the Louvre 200 years before Rumford wrote his treatise. Savat's fireplace achieved an amazing 30 to 40 percent greater efficiency than present-day American tract-home fireplaces. He surrounded the grate with a metal air chamber which had warm air outlets above the fire opening. He also supplied the fire with air from under the floor, thereby subsequently reducing room drafts and further improving combustion efficiency.

Few people realize that practically all of the technical features of Ben Franklin's 1742 Pennsylvania Stove were copied from earlier inventors. Savat's concept of pre-heated draft was employed by Franklin with little change in design. The descending flue was also copied as follows: smoke rose in front of a hollow metal back, passed over the top and down the opposite side. Finally, at the same level as the hearth, the smoke ascended the flue.

Wood Stoves

The stove sold today as "The Franklin Stove" contains none of the ingenious heat-saving features that were incorporated in the original models. The only real saving grace of modern Franklin Stoves is the closeable open-fire, which enables one to get a more efficient stove heat from a fireplace merely by closing a set of folding metal doors. According to The British Research Station, a closed fireplace unit is 50 percent more efficient than an open fire.

The English scientist, Dr. P. O. Rosin, has done much significant research on the aerodynamics of open fires. He built scale models of fireplaces using celluloid sheets to visually reveal the behavior of gaseous flow associated with open fires. Rosin disproved Rumford's contention that a plumb line from the center of the chimney flue should extend, uninterrupted, to the center of the fireplace fuel bed. Rosin pointed out that downdrafts have to be considered: eddies of smoke-laden air will back up into the room if they have an opportunity to collect in the smoke shelf area. Actually, after the experience of building scores of fireplaces over the past 20 years, I find that some eddies occur where the smoke shelf is minimal, as in Rumford's design. Rosin is correct: eddies will *not* occur where the shelf is eliminated entirely. But neither will they occur where the shelf is ample enough to allow the downdraft a free return trip up the chimney. Rosin's design includes a curved, free-flowing chimney breast and throat. The principle is good but impossible for the backyard mechanic to weld into place.

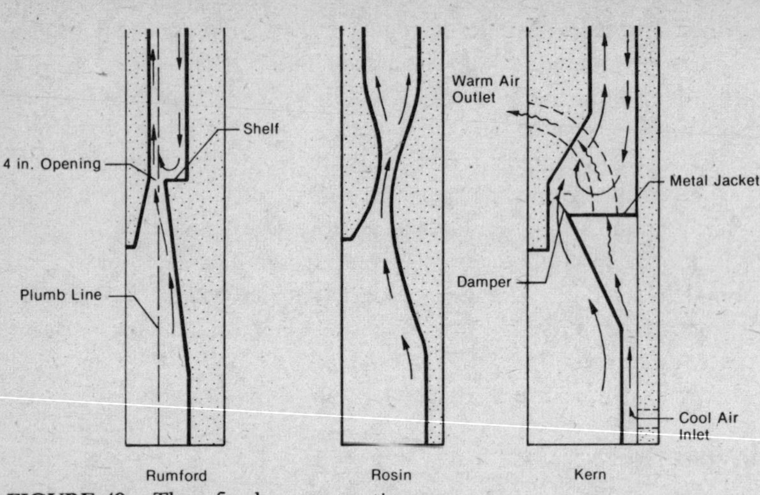

FIGURE 49. Three fireplace cross-sections.

WOOD COOKING RANGES

> A farm house should never be built without an ample open fireplace in its kitchen and other principally occupied rooms; and in all rooms where stoves are placed, and fires are daily required, the open Franklin should take the place of the close or airtight stove, unless extraordinary ventilation to such rooms be adopted also. The great charm of the farmer's winter evening is the open fireside, with its cheerful blaze and glowing embers; not wastefully expended, but giving out that genial warmth and comfort which, to those who are accustomed to its enjoyment, is a pleasure not made up by any invention whatever; and although the cooking stove or range be required which—in addition to the fireplace, we would always recommend, to lighten female labor—it can be so arranged as not to interfere with the enjoyment or convenience of the open fire.
>
> Lewis F. Allen, *Rural Architecture,* 1852.

Count Rumford was horrified at the dirty, inefficient, labor-making, fuel-devouring cooking range that adorned the English kitchen. Cooks at work over them, he said, ". . . looked and felt like buttered mummies," so great was the heat loss from the average range. Rumford solved the problem in 1800, but the solution has yet to trickle down to a 20th century biotechnic wood economy. The problem of heat emission was so well handled in Rumford's range that he later had to install a nearby fireplace grate to keep the cook warm! His principle was two-fold: produce heat only when needed, but not in excess; and, by insulating the range, use the heat before it is lost. His range was a massive brick affair. Cooking utensils were fitted into the masonry counter top and covered with insulating lids. Each utensil had its own separate fire source, grate, ash pit, and air regulator. His large ranges were built on a U-shape which certainly made a convenient work space for the cook. (See Figure 50.)

A hundred years elapsed between the time when Rumford made the first major breakthrough in range design and when a similarly efficient unit was produced in Sweden. During this interim millions of gross iron monsters were pawned off on an unsuspecting public. And, worst, possession of the ornate, grandmotherly iron range has become the personal dream today of nearly all back-to-the-landers, people who should know better! Like Rumford a hundred years before him, Dr. Gustaf Dalen had a chance to observe all the unnecessary chores associated with cooking. This one-time Swedish physicist and Nobel Prize winner was kept at home by blindness, and, while there, he made himself useful by developing what is perhaps the most efficient range ever built, the AGA. Again, like Rumford, Dalen's creation was so efficient that later models had to be modified to provide warmth for the cook. Pre-World War II models were advertised to run on about $10 worth of fuel—the cooking cost for a whole year! Dr. Dalen employed the same good-sense principles that Rumford laid down: produce heat only when needed and use it before it is lost. The same deep-well, insulated lids that Rumford proposed were incorporated in the AGA range.

Wood Furnaces

Emerson's house in Concord, Old Manse, was originally built by his grandfather. Typical of houses of that era, it contained numerous wood-burning fireplaces, heater stoves, a cooking range, and even a laundry stove. Each of these heating units had to be kindled and stoked, and each contained an ash pan and a flue pipe requiring constant maintenance. Finally, every wood-burning unit in the house established a unique environment of convection currents. That is, throughout the house there resulted many separate currents of air circulation with an ensuing heat loss through the multitude of chimney flues. Little wonder that coal-burning, warm-air, gravity-

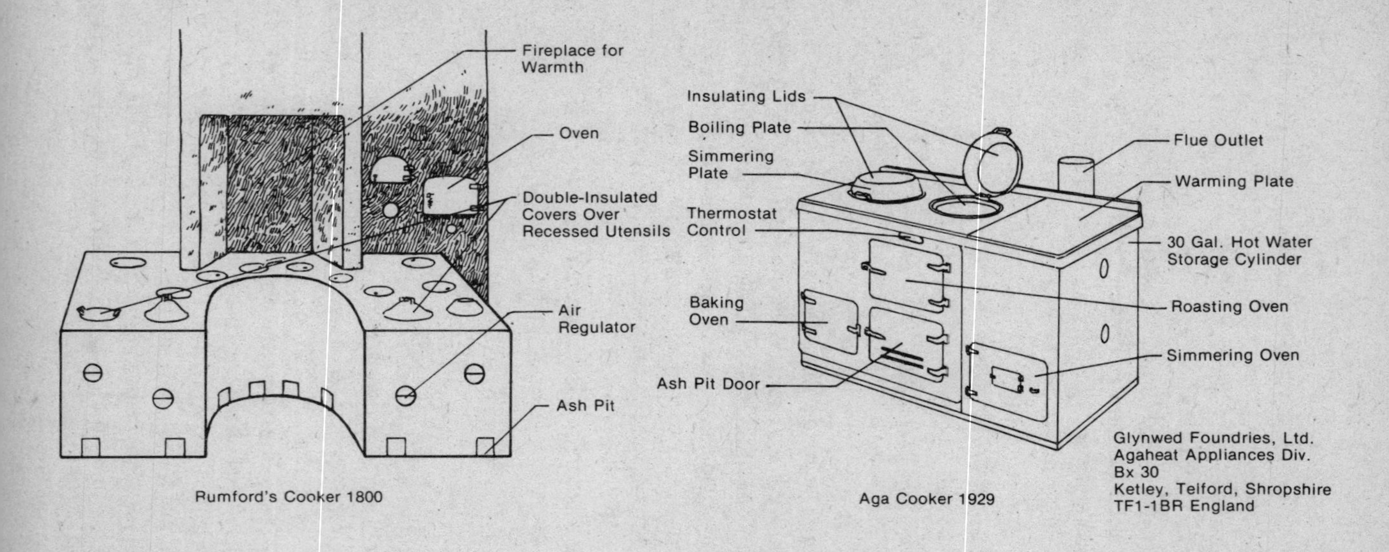

FIGURE 50. Two wood cookers.

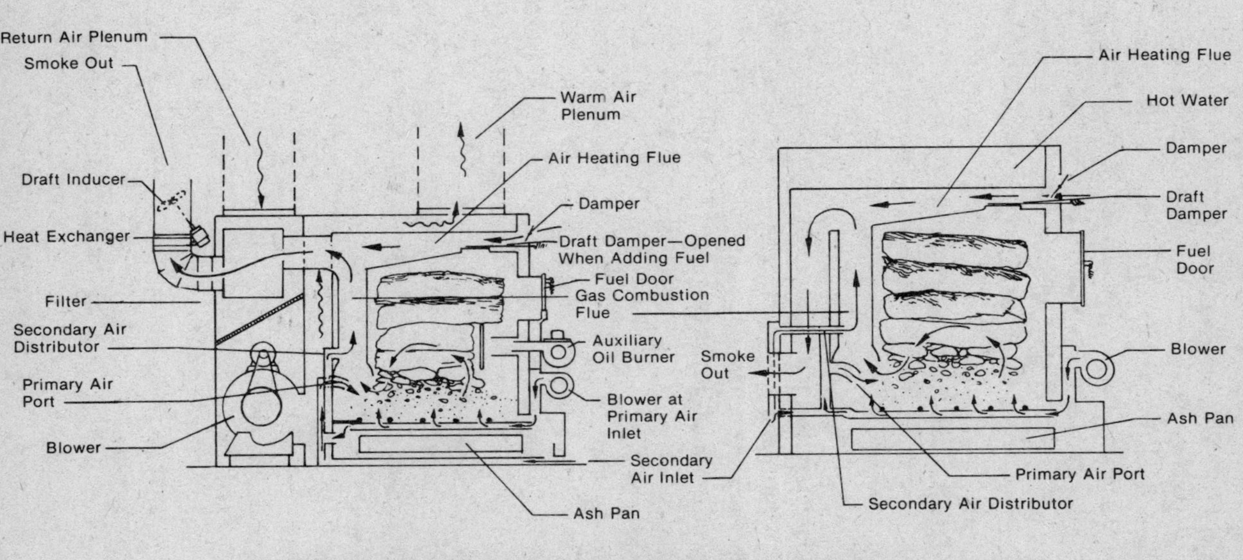

FIGURE 51. On the right, the Fuel-Master Wood-Burning Boiler; and on the left, the Fuel-Master Wood Furnace.

fed furnaces became the rage soon after Emerson's time. Located in the basement, this single facility provided heat and hot water for all the upper-level living and sleeping areas.

For years, an extremely efficient wood-burning furnace, the Riteway, was manufactured in this country. Until recently, however, the model was unavailable, but now, because of our recent energy shortages, the stove is again being manufactured.[2] Now called the Marco Fuel-Master, this furnace uses either hot air or hot water as a heat medium. Fans or pumps distribute the heat so that lower-level basement installation is not required. The hot-air furnace requires room-perimeter registers connected to the furnace plenum by insulated metal ducts. Hot water installation could be either in radiant pipe coils in concrete floors or in room-perimeter baseboard radiation coils.

The basic design features of both hot air or hot water furnaces were similar. First, and most important, the furnaces provided for a *complete fuel combustion.* This principle, described below, contrasts with the usual downdraft stoves, like the Ashley, in which fully 50 percent of the combustible gases leave the stove unburned, forming creosote and tars in the flue on the way out. Riteway and Fuel-Master furnaces and room heaters are equipped with a heavy cast-iron gas combustion flue located wholly inside the combustion chamber. This flue extends to the charcoal-burning level where wood gases accumulate. By providing several primary air jets over the charcoal bed, ignition of the wood gases is attained at the required 1200°F temperature. Preheated secondary air is then added to provide the necessary oxygen for complete combustion. Owners of Riteway and Fuel-Master stoves and furnaces report a fuel saving of up to 75 percent as a result of this complete combustion principle.

Both of these air-heating and water-heating furnaces have a wraparound jacket containing air or water. In addition, a heat exchanger is provided on the hot-air furnace. This consists of a large chamber through which hot smoke passes before reaching the flue. The main blower forces air across and around this hot smoke chamber, deriving additional warm air for the household in the process.

Speaking now of space-heating furnaces, Emerson would have heartily praised the efficiency of performance and the prudent design of the Riteway. He would, undoubtedly, flash on such special features as fire-brick lining in the heat chamber which gives the furnace a near-indestructible quality. The Riteway heater in our homestead home has been burning continuously throughout each of 15 consecutive winters, and it is still as good as the day we bought it. The self-sustaining, continuously burning feature of this heater makes it possible that only one fire need be built a season with only a few loadings every 24 hours to keep it going! An automatically maintained thermostat control, coupled with a heating unit that is completely airtight, makes fine control possible.

Building Your Own

This general background introduction to efficient wood heater design demonstrates how utterly wasteful the usual installation is. But it says more: by understanding the basic fundamentals of complete combustion we can design and build our own superior unit. I would wager the opinion that, in recent years, more wood heaters have been put together in small blacksmith and backyard welding shops than in all stove foundries combined. Cast-iron stoves were popular in Emerson's time because people liked the evenly distributed heat which they radiated. But cast-iron stoves cannot be made airtight, and, of course, they require factory techniques for their mass production. Cutting and welding metal is no longer the specialized craft it once was in our grandfather's time; simple arc and oxy-acetelyene equipment have replaced blacksmithing skills.

The traditional home-fabricated stove is made from a 55-gallon oil drum, laid horizontally. A pair of cradle legs made from bent angle iron is welded to the bottom side, and a fuel access door is cut out of the front end. The outlet flue is customarily located at the rear top of this lower-level barrel and connects to a second, upper-level, 55-gallon oil barrel which functions as both a heat chamber and an oven. The final flue outlet flows chimneyward at the upper back end of this heat chamber oven, although on more sophisticated barrel stoves the flue for this upper-level chamber is located above the front oven door and, thence, flows into the chimney. (See Figure 52.)

These stoves were so common in the Northwest territories at one time that they became known as Yukon stoves. To meet the popular demand for the Yukon barrel stove (also called the "belly stove"), manufacturers can supply conversion kits consisting of a cast-iron fuel door, legs, and a grate designed to fit the 55-gallon oil drum.[3]

This kind of heater installation is, at best, crude and inefficient, fuel-wise. But it does speak to one of the major shortcomings of American stoves; that is, to the lack of fuel capacity. Quite a number of 3-foot-long logs can be packed into the Yukon stove. The old-fashioned school stoves had this faculty, too. They were made from half a ton of cast-iron and could burn 24 hours with one loading. Logs 2 feet in diameter and 3 feet long could be packed into these stoves, which are now collectors' items. The epitome of limited fuel capacity is the traditional cast-iron—and the modern steel!—wood-burning stove.

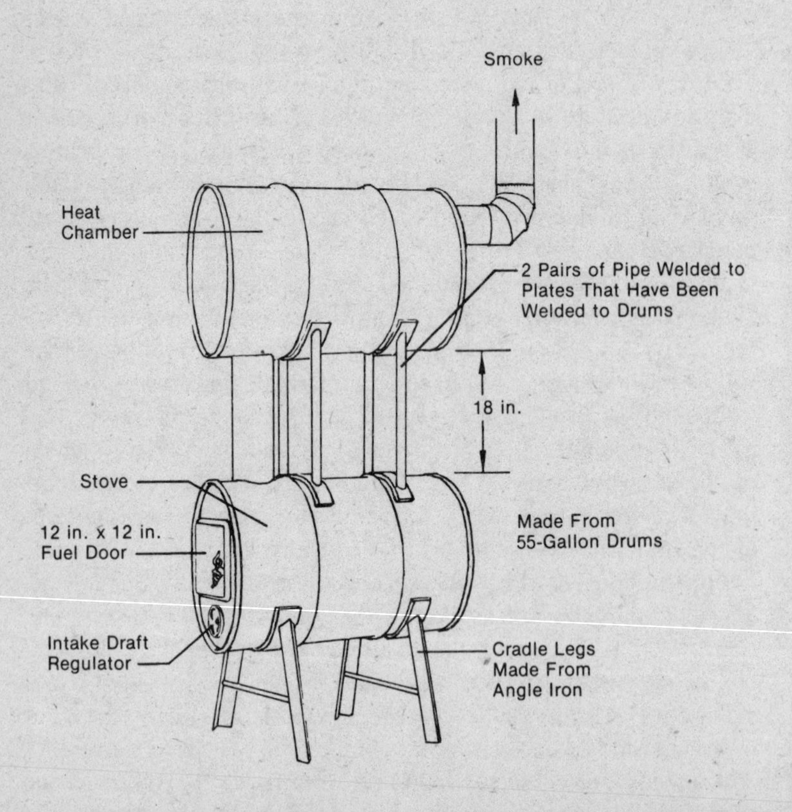

FIGURE 52. The common homemade stove, fashioned from two 55-gallon oil drums.

There is good reason to combine cooking and heating functions in one unit, especially for a compact, open-planned house such as my proposed Concord design. We already imagined how Emerson (or more likely his wife) must have trotted to keep half a dozen separate fires stoked on chilly days. And, remember, each fire required a separate flue, each hot metal flue contacting potentially flammable roofing material in the attic, compounding the risk of destructive fire.

Wood-to-flue joining substantially decreased the risk of fire danger if done properly, but the build-up of creosote and tar in the flue as a result of using an inefficient stove (or green wood) is another matter. Wood is one of the hottest burning fuels. Unburned gases condense in an uninsulated metal flue, forming caked layers of combustible creosote and tar. A spark will ignite this shaft of concentrated fuel, blasting flames out of a cherry-red chimney. This is why houses burn down in the middle of a cold winter.

For fire safety, as well as for economy, I recommend combining the cooking and stove heating functions into one facility. The flue from this dual-purpose unit should be clay tile, wrapped with fiberglass insulation and encased in the centrally located fireplace masonry.

Figure 53 illustrates some of the salient features of my proposed home-built cooking-heating unit. The combustion chamber consists of a 35-gallon oil drum set inside and welded to a 55-gallon oil drum outer shell. I chose oil drums because of their availability, their low cost, and their strong curvilinear construction. To provide a heat-absorbing cooking surface, a sheet of ¼-inch steel plate is welded to the top of the intersecting drums. Spaced bars of 1-inch reinforcing steel form the grate. Below the grate is welded a continuous length of 2-inch diameter steel pipe, perforated *at the bottom* with drilled holes to emit a continuous draft the full length of the fuel bed. Draft inlets are provided at the front of the stove for quick starting, and at the rear of the stove for the complete combustion of gases accumulating there. Both draft inlets are equipped with positive, fine-adjustment controls. The rear draft is supplied from under-the-floor air space, thereby creating a full-circle, no-draft convection current. During colder temperatures a front-mounted squirrel cage fan (almost silent) is activated to help circulate jacketed hot air into the room. A second 55-gallon drum encases the oven at a

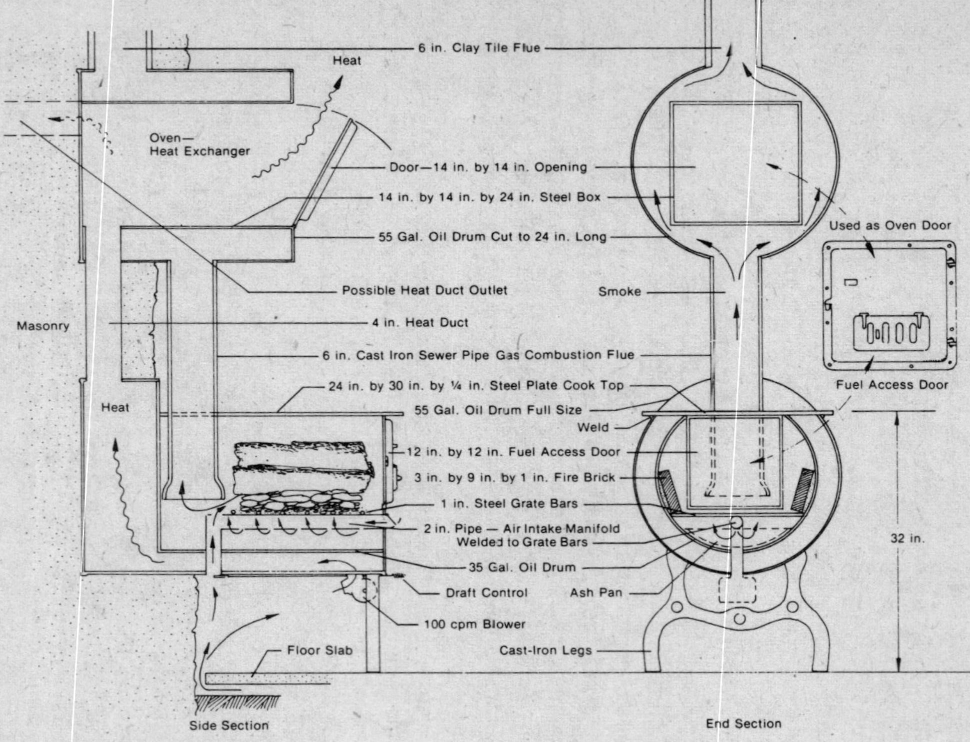

FIGURE 53. Kitchen heater-cooking range-oven.

convenient level above the cooking surface of the lower-level drum. Both drums are cantilevered out of the masonry wall. Hot air flows from the central chamber of this upper-level drum through a metal-boxed heat exchanger, and from there into the room.

One of the most unique—and valuable—features of the Riteway or Fuel-Master stove is the gas combustion flue. The principle certainly should be included in our home-made version. Due to the high temperatures incurred at the point of gas combustion, a heat-resistant material, such as cast-iron, should be used. The ideal flue is made from a length of 6-inch sewer pipe with the hub down to augment the combustion capacity. It is connected directly to the heat exchanger.

Our total winter-time hot water needs are met because a thoughtful Riteway engineer provided space in the firebrick lining of the combustion chamber for a metal, water-heating collector. This one simple installation has, doubtless, saved enough electricity or gas to repay the original cost of the heater many times over through its uncomplicated, trouble-free, 15-year performance for my family. At one time Superior Fireplace Company, makers of the Heatform fireplace liner, offered a water-heating coil that fastened onto the metal fireback. Drawing hot water from heater and fireplace is an excellent idea and should certainly be included in the new Concord dwelling. Heated water naturally thermosiphons into an upper-level storage tank. In summer months a solar collector provides hot water. The solar unit, the fireplace, and the stove units are all tied into the same storage tank.

Lastly, if a sauna is desired (Emerson would have dug it!), an additional hot water supply can be provided with a Nippa woodburning sauna heater.[4] They come equipped with either a 20-gallon water jacket or a heat collector which can be connected to an insulated water storage tank. In the new Concord house I have located the sauna and bathing facilities on the upper level of a compost-privy core. When the room is not used as a sauna, the stove provides heated water and comfortable warmth. (See Figure 42.)

Woodlot Management

Wood-fueled heating and cooking is an especially attractive energy alternative when the homeowner has at his or her disposal some

—even just a few—acres of woodland to manage. An average-growth woodland managed on a sustained-yield basis will supply about one cord of wood per year per acre of woodland. The wood removed is in large part in the form of thinnings and prunings, although the term, "sustained yield," infers that the amount of wood removed each year is equal to the amount that grows back. Seedlings are nurtured or planted wherever mature trees are removed. In the instance where the homeowner has unplanted acres available, he or she can do no better than to plant a woodlot in the form of a windbreak.

The term "shelterbelt" is often used interchangeably with "windbreak," but a distinction should be made between the two terms. A windbreak is a protective planting around one's house and garden, whereas a shelterbelt is a long, planted barrier protecting large fields. During the Great Depression, the Prairie States' Shelterbelt Project under the auspices of the U.S. Forest Service was implemented, and over 30,000 separate shelterbelts were planted, reaching a total length of about 19,000 miles. Most of what we now know about planting trees for wind protection and for fuel supply has come out of the extensive research and experimentation done during this project.

SHELTERBELTS

Towards the end of the Shelterbelt Project (in 1942), plans were developed for an optimum, "standard shelterbelt" proposal. These plans included provision for planting varieties contributing early protection and for planting semi-permanent and permanent varieties. Rows of permanent, dense, low-growing shrubs and conifers were planted on the windward side of this "standard" belt. Rapidly growing, deciduous species were placed in the center of the belt, and, on the leeward side, longer-lived varieties were planted to provide early wind protection. (See Table 9.) A few rows of a shrubby species were also planted at the far end of the leeward side of the shelterbelt, bringing the total number of rows to about ten. (See Figure 54.) Shelterbelts of less than six rows failed to develop a desirable forest ecology.

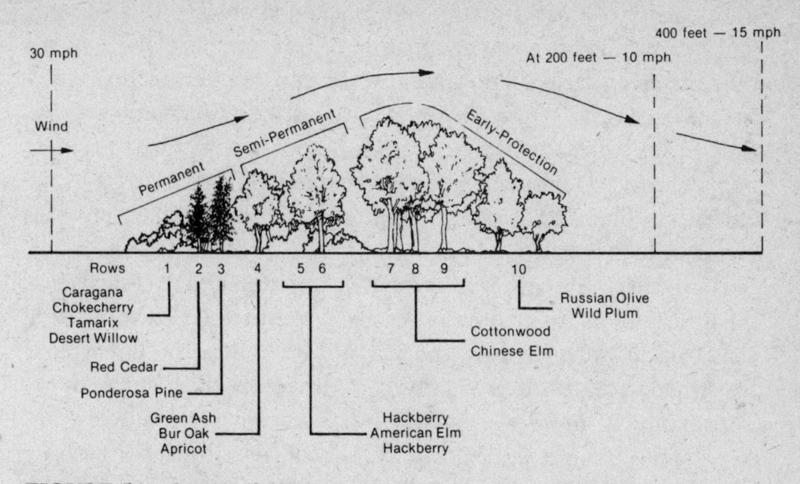

FIGURE 54. Suggested planting of tree varieties in a shelterbelt.

During the 6½ years of the Shelterbelt Project, it was found that the amount of growth of a tree's height increased from northern climates southward. In North Dakota the average growth in height for this period was 16 feet; in Nebraska it was 20 feet; and in Texas it was 24 feet.

Row spacing of 10 feet apart was established between the trees. Distance separating trees in each row was not over 8 feet, and shrubs were not over 4 feet distant from each other. During the six and one-half year project, the crowns of the rapidly growing species closed in fully, and a forest floor mulch an inch thick was formed during this same period.

Primarily as a result of the U.S. Forest Service efforts, there has

TABLE 9. Relative Diameter Growth

RAPID	MODERATE	SLOW
European Larch	Douglas Fir	Cedar
Loblolly Pine	Ponderosa Pine	Hemlock
Aspen	Redwood	Longleaf Pine
Black Locust	Spruce	Beech
Cotton Wood	Black Walnut	Oak-Black & White
Willow	Elm	Sugar Maple

developed a distinct but thorough science of shelterbelt planning. This science involves the designation of tree varieties, planting layout, and tree spacing. Poplars, for instance, are often used as windbreaks because they are able to withstand high-wind pressure, but they are also great soil robbers. So it becomes prudent to alternate poplars with the cultivation of a complimentary variety, such as the alder tree. The alder is one of the few non-legumes which provides nitrogen to the soil.

It is likewise important to combine coniferous and deciduous tree plantings. Conifers planted on eroded, compacted, humus-lacking soils will, in time, improve the soil and its humus content so that hardwood trees planted nearby can thrive on the increased soil nutriments. A mixed woodland is much less subject to insect damage, mainly because of the greater variety of trees. Insects are usually attracted to specific tree varieties. In a well-planned woodlot the extent of insect damage is limited because of the small number of any one type of tree in any one area. An example of woodlot mismanagement is the fatal Dutch elm disease which was introduced by the elm bark beetle and was responsible for killing mono-planted, roadside elm trees all over the East. Concentrated likeness invited and caused widespread destruction.

In addition, a mixed woodlot produces a mixed, well-balanced layer of leaves on the forest floor. This loose, crumbly layer of mulch is certainly superior to a dense, impermeable layer of a single variety of needles or leaves.

Someone once discovered that trees have three different shaped root systems; a *spear* shape found in the oak which taps minerals from great depths; a *heart* shape found in the birch which lifts huge quantities of water to its top; and a *flat* shape designed for the support required by such trees as the giant Sitka spruce. Obviously, a plantation of a single species of tree with identical root systems would offer fierce competition for food, moisture, and support at root level. This is another good reason why one should make certain that his or her woodland contains a mixed variety of trees, which has wide variance in age and in growth patterns. A dense monoculture of conifers may successfully break the force of prevailing winter winds, but, at the same time, they may obstruct the flow of cold air, impeding natural air drainage. The winter-bared branches of deciduous trees will not obstruct this important cold air movement.

WINDBREAKS

By far the most effective arrangement for a windbreak near one's dwelling is in the form of an "L," with the point of this "L"-planting directed at the prevailing winter winds. (See Figure 55.) This layout is best both for preventing evaporation of soil moisture, thus helping to raise soil temperatures, and for catching and preventing the drifting of snow around walks and buildings where it could otherwise be an annoying problem. (I should point out, however, that soil mois-

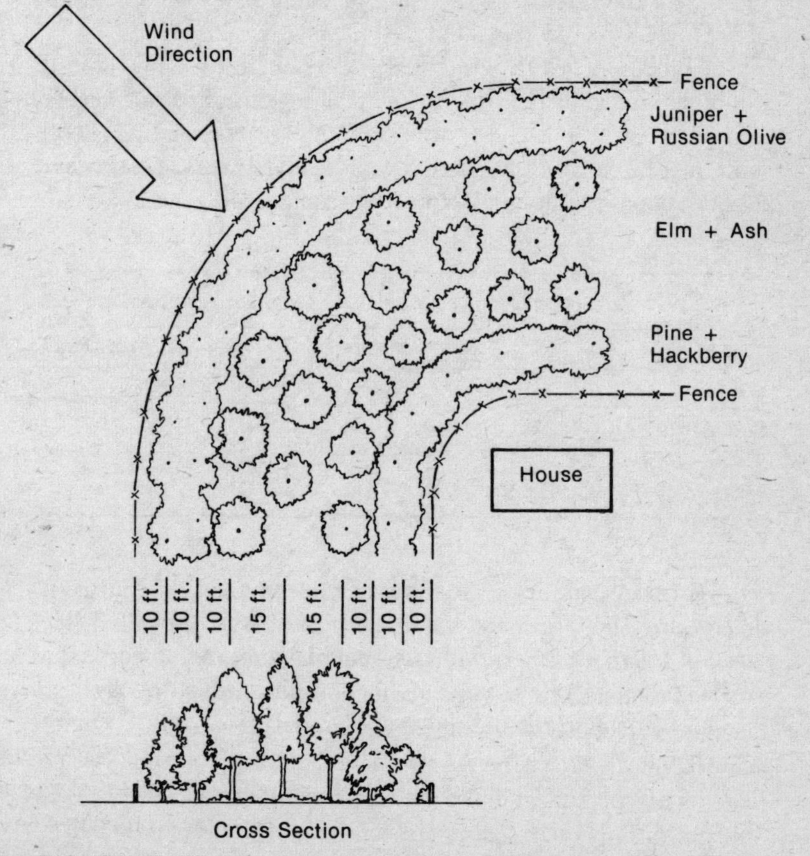

FIGURE 55. Ground plan for windbreak designed for Plains States.

ture for crops can be conserved if snow is held in the shade of shelter trees where it melts slowly in the spring.)

One of the more valuable experiments which evolved from the Prairie States' Shelterbelt Project subsequently resulted in remarkable savings in heating fuel from planted windbreaks. In these Nebraska experiments exact fuel consumption was recorded in two identical test houses. One of these houses was exposed to the winds and one was protected by a nominal windbreak. Both houses were maintained at a constant inside temperature of 70°F. The house having windbreak protection required *30 percent less fuel* than the exposed dwelling. It was also found that during a mild winter, animals in a tree-protected yard gained 35 percent more weight than those in an open area. It is, therefore, apparent that a windbreak not only provides fuel to burn, but fuel savings as well.

The process of selecting the best species of trees for a windbreak can be simplified by referring to charts of regions of natural vegetation prepared by the U.S. Forest Service. The species you choose must, first be climatically suited to your regional area. Selection also depends upon specific site factors. For example, the inherent capacity of a tree species to withstand shade becomes a major factor in determining the variety and the numbers of trees to be planted. The range of tolerance is very wide, as Table 10 indicates. Sugar maple seedlings require ony 2 percent full sunshine, while lobolly pine seedlings require nearly full sunlight to grow satisfactorily. Seedlings are generally more tolerant of shade than mature trees, especially if they are grown in a good exposure and on good soil. Shade-givers, growing near seedlings, can provide valuable physical protection to young trees.

TABLE 10. Relative Tolerance to Shade of Common Woodland Trees

TOLERANT	INTERMEDIATE	INTOLERANT
Cedar	Douglas Fir	Pine
Hemlock	Ash	Larch
Redwood	Birch	Aspen
Spruce	Chestnut	Black Walnut
Beech	Elm	Hickories
Maple	Oak	Willow

Calculating Your Wood Fuel Needs

Fuel wood is usually measured by the cord. The standard cord is a pile 8 feet long, 4 feet wide and 4 feet high. Comprehensive labor studies in New York State found that an experienced woodchopper can fell, skid, saw, and split 2 cords per day. The study was made before chain saws came into common usage, so you can estimate that an inexperienced person using a power chain saw could cut this much a day. However, proper management of a windbreak could perhaps double the amount of time required to extract a cord of wood, for at least one-half the harvest should properly consist of small limb thinnings and prunings and it takes more time to gather a cord of these thinner pieces than it does regular logs.

Even at a cutting rate of one cord per day, a week's labor, spread over the winter months, should be sufficient to supply the fuel needs for a whole heating-cooking season in a cold, temperate climate. Of course, this estimate must be qualified to take into account the species of wood selected and its moisture content at the time of burning. Although wood substances are all fundamentally the same in chemical composition, some varieties contain more resins, oils, and gums than other varieties of wood and therefore have a higher calorific value. (See Table 11.) Calorific value represents the total amount of heat given off by the wood burned, and this value is greatly influenced by the amount of moisture in the wood. The British Thermal Unit (Btu) is a common measurement of heat quantity, taking one Btu to raise one pound of water 1°F.

Table 11. Heating Value per Cord of Different Woods

WOODS	WEIGHT AIR DRY	AVAILABLE HEAT MILLION BTU AIR-DRY
Ash	3440	20.0
Aspen	2160	12.5
Beech	3760	21.8
Birch	3680	21.3
Douglas Fir	2400	18.0
Elm	2900	17.2
Hickory	4240	24.6
Maple	3200	18.6
Oak, Red	3680	21.3
Oak White	3920	22.7
Pine-Eastern White	2080	13.3
Pine-Southern Yellow	2600	20.5

The need for a large storage area to properly air-dry firewood has been one of the major objections to using wood as fuel. A simple, covered storage area is all that is required to properly air-dry firewood. In areas having a high humidity, a full year of storage may be necessary to "cure" one's wood supply for fuel use. As Table 12 illustrates, the moisture content of the wood to be used dramatically affects the heat available from that wood. Air-dried wood with about a 20 percent moisture content will supply about five times more Btu's than will green wood.

TABLE 12. Heat Available From 1 Pound of Wood

PERCENTAGE OF MOISTURE CONTENT	AVAILABLE HEAT BTU
Ovendry— 0	7100
5	6700
10	6300
15	5800
Air-dry—20	5400
25	4800
30	4500
35	4000
40	3600
45	3400
50	2900
55	2500
60	2100
65	1500
70	1100

In Emerson's time the per capita consumption of wood fuel reached an all-time peak with an average yearly use of four and one-half cords per person. Over 150 million cords of wood were burned in 1880; in 1952, only one-third this amount was used. Very little cordwood cut today is seriously used for heating and cooking. It is, rather, limited to the grossly inefficient but, nevertheless, esthetically satisfying fireplace.

But esthetics need *not* be sacrificed to efficient function. I hope that I have succeeded in demonstrating here that highly efficient heating and cooking facilities *can* be installed by anyone who is eager and willing, and that they can be built in conjunction with an array of building design features that take advantage of solar heat and insulation and no-draft, warm air circulation. It is indeed gratifying to discover that the time and energy that is put into woodlot management and wood chopping is rewarded by the practicalities and pleasure one gets from efficient fireplaces, wood stoves, and furnaces. An acknowledgement of this gratification is reflected in the observation of Emerson's neighbor, Thoreau, who remarked that, "Wood heats you twice; once when you cut it, and once again when you burn it." I would add, ". . . and yet once again as one's wood fire is tended."

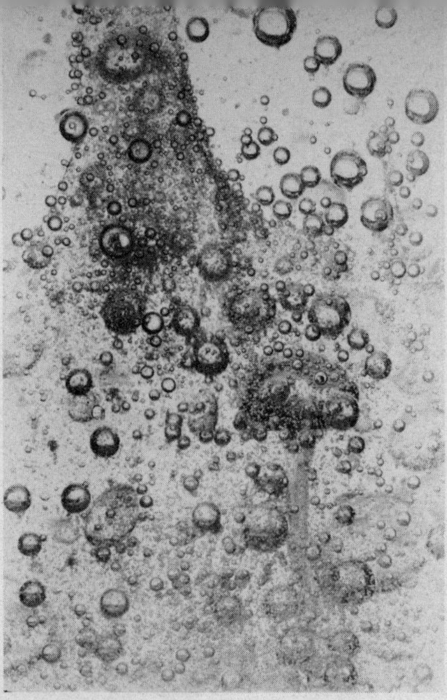

4 METHANE POWER

Methane Gas Digesters for Fuel and Fertilizer

The New Alchemy Institute West

Recently, attention has turned to methane digesters as a source of fuel gas and fertilizer. The interest is understandable in view of the mounting shortages of energy sources (whether real or political) and the increasing desire of many to develop a more self-sufficient pattern of living . . . especially in rural areas.

However, much of the information concerning digesters and digester systems has been misleading and overly complex. It has avoided basic questions such as: how much raw organic material can be expected from the plant or animal wastes available? How much gas will they produce? What kind and size of digester should be built (so that it suits the needs and resources of whoever builds it)? And how is the digester started?

The answers to these questions aren't that difficult, and we have found that productive digester operations can be built and maintained by knowing some things about the biology of digestion and the properties of the raw materials going into the digester.

In this section we would like to present a general background of the raw materials and processes of digestion and discuss some preliminary ideas for using methane gas and sludge.

Background

When organic material decays it yields useful by-products. The kind of by-product depends on the conditions under which decay takes place. Decay can be *aerobic* (with oxygen) or *anaerobic* (without oxygen). Any kind of organic matter can be broken down either way, but the end products will be quite different. (See Figure 56.)

It is possible to mimic and hasten the natural anaerobic process by putting organic wastes (manure and vegetable matter) into in-

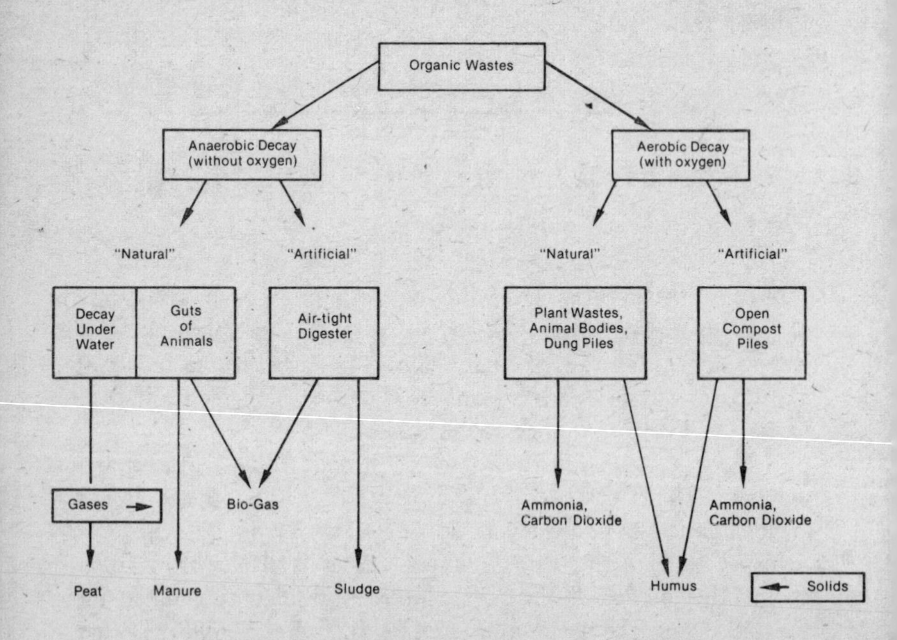

FIGURE 56. End products of organic decay.

sulated, airtight containers called *digesters*. Digesters are of two types: (1) *Batch-load digesters* which are filled all at once, sealed, and emptied when the raw material has stopped producing gas; and (2) *Continuous-load digesters* which are fed a little, regularly, so that gas and fertilizer are produced continuously.

The digester is fed with a mixture of water and wastes, called "slurry." Inside the digester, each daily load of fresh slurry flows in one end and displaces the previous day's load which bacteria and other microbes have already started to digest. Each load progresses down the length of the digester to a point where the methane bacteria are active. At this point large bubbles force their way to the surface where the gas accumulates. The gas is very similar to natural gas and can be burned directly for heat and light, stored for future use, or compressed to power heat engines.

Digestion gradually slows down toward the outlet end of the digester and the residue begins to stratify into distinct layers (see Figure 57): *Sand and Inorganic Materials* at the bottom.

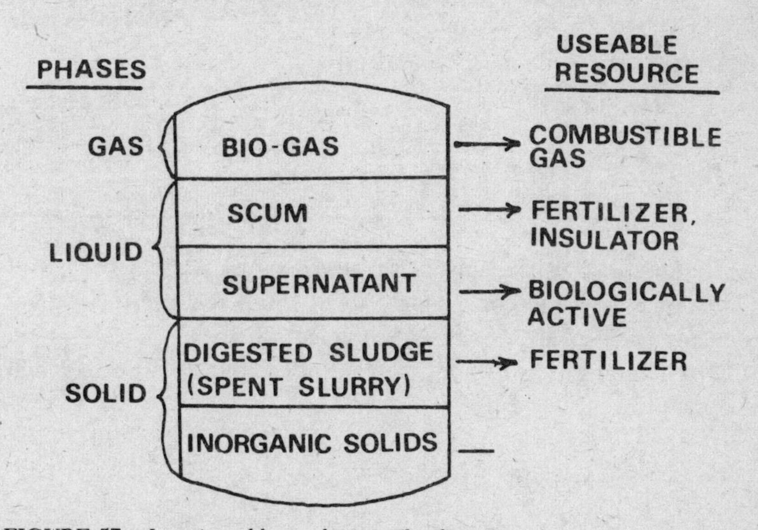

FIGURE 57. Layering of by-products in the digester.

Sludge, the spent solids of the original manure reduced to about 40 percent of the volume it occupied in the raw state. Liquid or dry sludge makes an excellent fertilizer for crops and pond cultures.

Supernatant, the spent liquids of the original slurry. Note that the fertilizing value of the liquid is as great as sludge, since the dissolved solids remain.

Scum, a mixture of coarse fibrous material, released from the raw manure, gas, and liquid. The accumulation and removal of scum is one of the most serious problems with digesters. In moderate amounts, scum can act as an insulation, but in large amounts it can virtually shut down a digester.

TABLE 13. Total Fuel Value of U.S. Methane Resources Supplied by Digestion of Readily Collectable, Dry, Ash-Free Organic Wastes.

For perspective, consider the total fuel value of methane that could be produced from the available organic wastes in the United States.

Fuel Value of U.S. Methane Resources (ref. 1)	
A. Organic wastes in U.S./year	2 billion tons (wet weight) 800 million tons (dry weight)
B. Dry organic waste readily collectable	136.3 million tons
C. Methane available from "B" (@10,000 ft^3/ton)*	1.36 trillion ft^3/year
D. Fuel value of methane from "C" (1000 Btu/ft^3)	1,360 trillion Btu/yr
Fuel Consumption of U.S. Farm Equipment (ref. 2)	
A. Total gasoline consumed (1965)	7 billion gallons/year
B. Total energy consumed by "A" (1 gallon gasoline = 135,000 Btu)	945 trillion Btu/year
Total U.S. Natural Gas Consumption (1970)	19,000 trillion Btu
Total U.S. Energy Consumption (1970)	64,000 trillion Btu

*Urban refuse; higher figure for manure and agricultural wastes.

So, speaking generally, methane gas converted from *easily available* organic wastes could supply about 150 percent of the gasoline energy used by all U.S. farm equipment (1965), 7 percent of the 1970 natural gas energy, or 2 percent of the total 1970 U.S. energy demands.

METHANE-GAS PLANT: SYNERGY AT WORK

When we consider digesters on a homestead scale, there are two general questions to ask: (1) with the organic wastes and resources at hand, what kind of digester should be built, and how big should it be? and (2) what is the best way of using the gas and sludge produced to satisfy the energy needs of the people involved? (whether the sludge should be used to fertilize crops, fish or algae ponds, and whether the gas should be used directly for heat and light, or stored, or fed back to the digester to heat it, etc., see Figure 58).

The first question involves the digester itself, which is the heart of a whole energy system. The second question is synergistic: you can choose which products are to be generated by digestion and you can choose to use them or feed them back to the digester, creating an almost endless cycle if you wish. (See Figure 59.)

The model in Figure 59 is idealized from oriental aquaculture systems and other ideas, both old and new. A single pathway can be developed exclusively (have your digester produce only sludge to feed an algae pond) or you can develop the potential synergy (many possible systems working together as an integrated whole, see Figure 60).

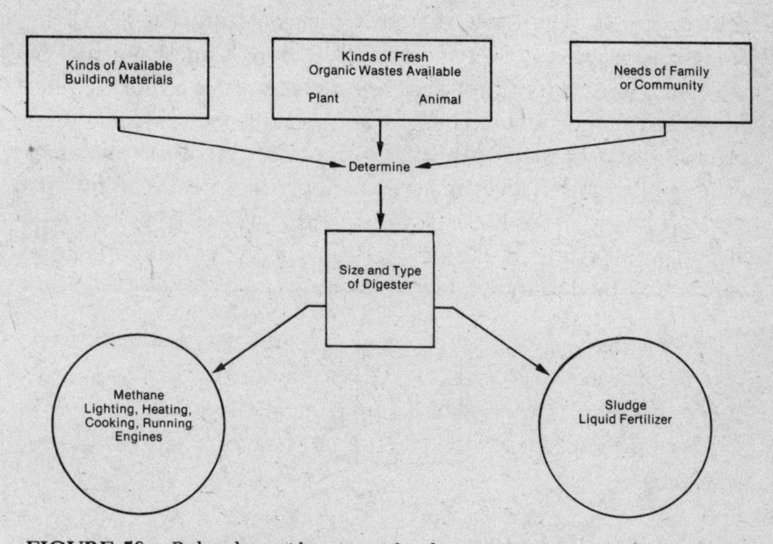

FIGURE 58. Related considerations of a digester operation.

The small farmer or rural homesteader can take a step toward ecological self-sufficiency by producing some of his fuel and fertilizer needs using a digester to convert local wastes. Total dependence on conventional fuels, especially in rural areas, is likely to become a serious handicap in the years to come as reserve shortages and specialized technologies hike the costs of fossil and nuclear fuels. But by producing energy from local resources, it is possible to be partially freed from remote sources of increasingly expensive fuel supplies.

History

In nature, anaerobic decay is probably one of the earth's oldest processes for decomposing wastes. Organic material covered by a pool of warm water will first turn acid and smell rank, then slowly —over about six months—will turn alkali. The methane bacteria, always present, will take over and decompose it, and gas bubbles will rise to the surface.

Anaerobic decay is one of the few natural processes that hasn't been fully exploited until recent times. Pasteur once discussed the possibilities of methane production from farmyard manure. And (according to a report issued from China on April 26, 1960) the Chinese have used "covered lagoons" to supply methane fuel to communes and factories for decades. But the first attempt to build a digester to produce methane gas from organic wastes (cow dung) appears to have been in Bombay, India in 1900. At about this time, sewage plants started digesting sewage sludge in order to improve its quality. The success of these plants started a mass of laboratory and small-scale experiments during the 20's and 30's (many of them summarized by Acharya, ref. 3).

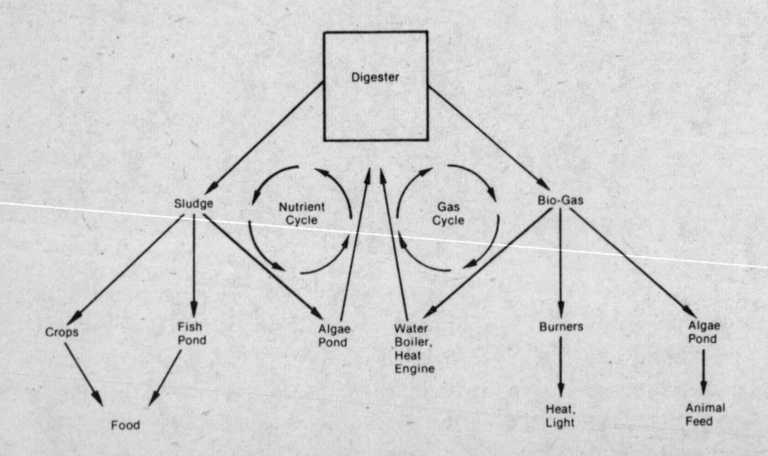

FIGURE 59. The closed nutrient system of a complete digester operation.

During World War II, the shortage of fuel in Germany led to the development of methane plants in rural areas, where the gas was used to power tractors. The idea spread into Western Europe, but died out when fossil fuels once again became available (although, today, many farmers in France and Germany continue to use home digesters to produce their own methane fuel gas).

Currently the focus of organic digester/bio-gas research is primarily in India. India's impetus has been the overwhelming need of a developing country to raise the standard of living of the rural poor. Cows in India produce over 800 million tons of manure per year; over half of this is burned for fuel and thus lost as a much needed crop fertilizer.[4] The problem of how to obtain cheap fuel *and* fertilizer at a local level led to several studies by the Indian Agricultural Research Institute in the 1940's to determine the basic chemistry of anaerobic decay. In the 1950's, simple digester models were developed which were suitable for village homes. These early models established clearly that bio-gas plants could: (1) provide light and heat in rural villages, eliminating the need to import fuel, to burn cow dung, or to deforest land; (2) could provide a rich fertilizer from the digested wastes; and (3) could improve health conditions by providing airtight digester containers, thus reducing disease borne by exposed dung.

More ambitious designs were tested by the Planning Research and Action Institute in the late 1950's. Successes led to the start of the Gobar Gas Research Station at Ajitmal where, with practical experience from the Khadi and Village Industries Commission, two important pamphlets[5, 6] were published on the design of village and homestead "bio-gas" plants in India.

In America, where the problem is waste disposal, rather than waste use, organic digesters have been limited to sewage treatment plants.[7, 8] In some cases sludge is recycled on land or sold as fertilizer,[9, 10] and methane gas is used to power generators and pumps in the treatment plants.[11] The Hyperion sewage treatment plant in Los Angeles generates enough methane from its primary treatment alone to power its 24-2,000-hp. diesel engines. Usually, however, both sludge and gas are still regarded as waste problems.

Much information on digestion and small–scale digester operations comes from experiences in India, Western Europe, and South Africa, and journals such as: *Compost Science, Water Sewage Work,*

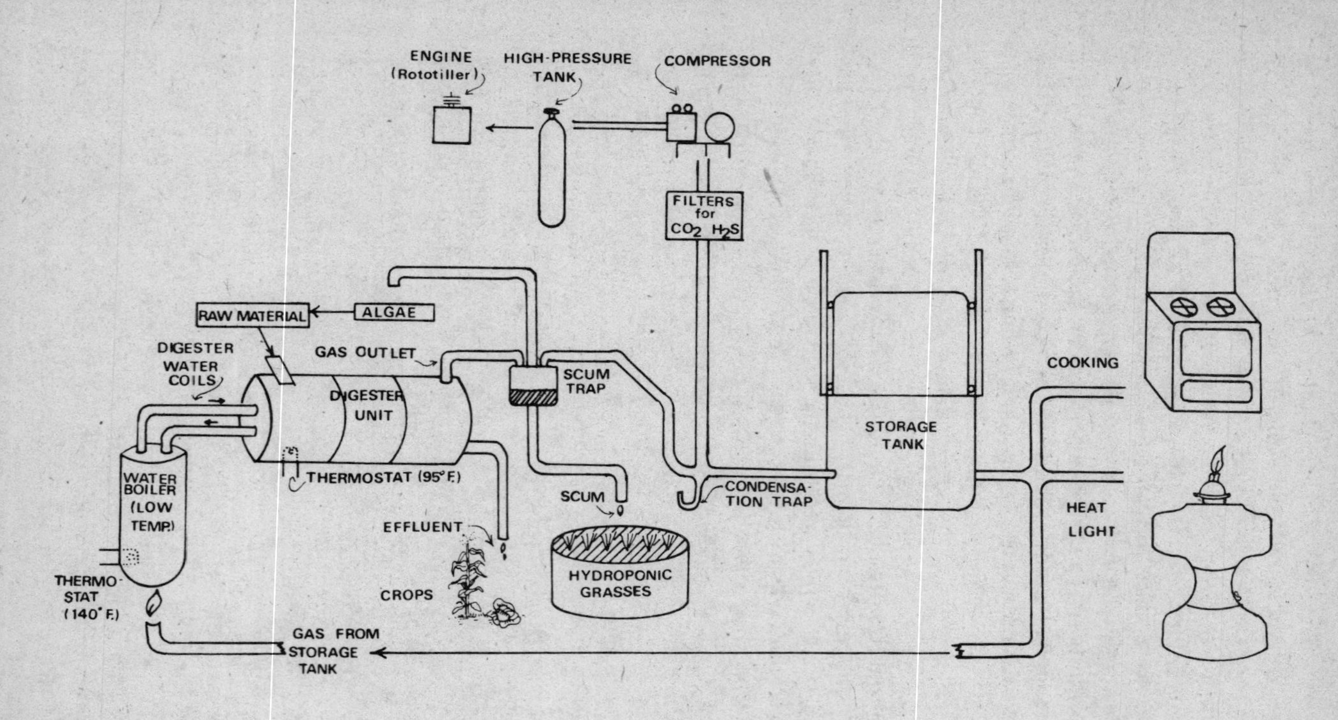

FIGURE 60. Integrated organic digester operation (using 50-gallon drums for digester).

Soils and Fertilizer, Waste Engineering, Sewage and Industrial Wastes and recent publications of the U.S. Environmental Protection Agency and Solid Waste Conferences. An excellent book to learn from is called: *Manual of Instruction for Sewage Plant Operators,* put out by the New York State Department of Health.

A great deal of information can be found in pre-World War II sewage journals, especially *Sewage Works Journal.* After World War II, as with most other kinds of science and technology, waste treatment research became a victim of the trend to make machines ever bigger, and information increasingly incomprehensible.

Biology of Digestion

Bio-Succession In The Digester

Perhaps the most important thing to remember is that digestion is a biological process. The "anaerobic" bacteria responsible for digestion can't survive with even the slightest trace of oxygen. So, because of the oxygen in the manure mixture which is fed to the digester, there is a long period after loading before actual digestion takes place. During this initial "aerobic" period, traces of oxygen are used up by oxygen-loving bacteria, and large amounts of carbon dioxide (CO_2) are released.

When oxygen disappears, the digestion process can begin. This process involves a series of reactions by several kinds of anaerobic bacteria feeding on the raw organic matter. As different kinds of these bacteria become active, the by-products of the first kind of bacteria provide the food for the rest. (See Figure 61.)

In the first stages of digestion, organic material which is digestible (fats, proteins, and most starches) are broken down by acid-producing bacteria into simple compounds. The acid bacteria are capable of rapid reproduction and are not very sensitive to changes in their environment. Their role is to excrete enzymes, liquefy the raw materials, and convert the complex materials into simpler substances (especially volatile acids, which are low molecular weight

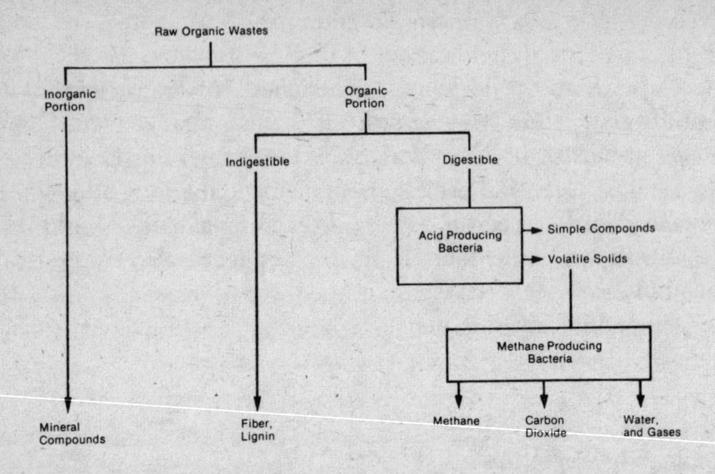

FIGURE 61. The biological breakdown of wastes in the digester.

organic acids—see **Raw Materials**). The most important volatile acid is acetic acid (table vinegar is dilute acetic acid), a very common by-product of all fat, starch, and protein digestion. About 70 percent of the methane produced during fermentation comes from acetic acid.[12]

Once the raw material has been liquefied by the acid-producing bacteria, methane-producing bacteria convert the volatile acids into methane gas. Unlike the acid bacteria, methane bacteria reproduce slowly and are very sensitive to changes in the conditions of their environment. (More information on the biology of methane fermentation can be found in ref. 13 and 14.)

Biologically, then, successful digestion depends upon achieving and (for continuous-load digesters) maintaining a balance between those bacteria which produce organic acids and those bacteria which produce methane gas from the organic acids. This balance is achieved by a regular feeding with enough liquid (see **Feeding**) and by the proper pH, temperature, and the quality of raw materials in the digester.

pH AND THE WELL-BUFFERED DIGESTER

To measure the acid or alkaline condition of a material, the symbol "pH" is used. A neutral solution has pH = 7; an acid solution has pH below 7; and an alkaline solution has pH above 7. The pH has a profound effect on biological activity, and the maintenance of a stable pH is essential to all life. Most living processes take place in the range of pH 5 to 9. The pH requirements of a digester are more strict (pH 7.5–8.5, see Figure 62).

During the initial acid phase of digestion, which may last about two weeks, the pH may drop to 6 or lower, while a great deal of CO_2 is given off. This is followed by about three months of a slow decrease in acidity during which time volatile acids and nitrogen compounds are digested and ammonia compounds are formed (this ammonia becomes important when we consider the fertilizer value of sludge).

As digestion proceeds, less CO_2 and more methane is pro-

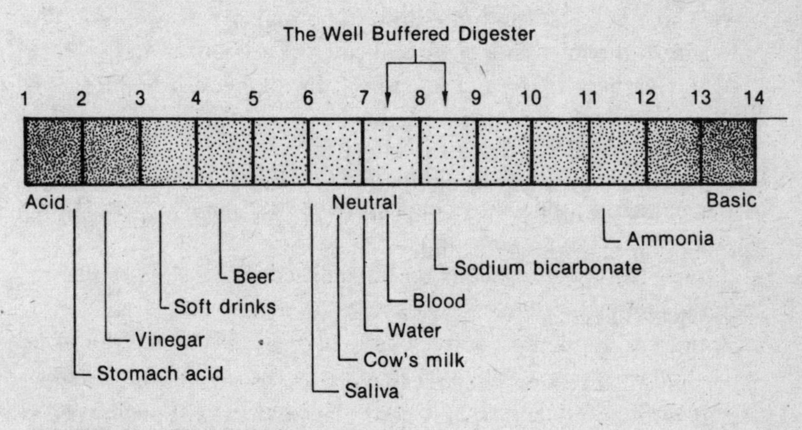

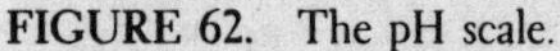

FIGURE 62. The pH scale.

duced, and the pH rises slowly to about 7. As the mixture becomes less acid, methane fermentation takes over. The pH then rises above the neutral point (pH = 7), to between pH 7.5 and 8.5. After this point, the mixture becomes well buffered; that is, even when large amounts of acid or alkali are added, the mixture will adjust to stabilize itself at pH 7.5 to 8.5.

Once the mixture has become well buffered, it is possible to add small amounts of raw material periodically and maintain a constant supply of gas and sludge (continuous-load digesters). If you don't feed a digester regularly (batch-load digesters), enzymes begin to accumulate, organic solids become exhausted and methane production ceases.

After digestion has stabilized, the pH should remain around 8.0 to 8.5. The ideal pH values of effluent in sewage treatment plants is 7 to 7.5, and these values are usually given as the best pH range for digesters in general. From our experience, a slightly more alkaline mixture is best for digesters using raw animal or plant wastes.

You can measure the pH of your digester with litmus or pH paper which can be bought at most drug stores. Dip the pH paper into the effluent as it is drawn off. Litmus paper turns red in acid solutions (pH 1 to 7) and blue in alkaline solutions (pH 7 to 14). You can get more precise measurements using pH paper which changes colors within a narrow range of pH values.

If the pH in the continuous-load digester becomes too acidic (see Table 14), you can bring it up to normal again by adding fresh effluent to the inlet end, or by reducing the amount of raw material fed to the digester, or as a last resort, by adding a little ammonia. If the effluent becomes too alkaline, a great deal of CO_2 will be produced, which will have the effect of making the mixture more acidic, thus correcting itself. Patience is the best "cure" in both cases. NEVER add acid to your digester. This will only increase the production of hydrogen sulfide.

TEMPERATURE

for the digesting bacteria to work at the greatest efficiency, a temperature of 95°F (36°C) is best. Gas production can proceed in two ranges of temperature: 85° to 105° and 120° to 140°F. Different sets of acid-producing and methane bacteria thrive in each of these different ranges. Those active in the higher range are called heat-loving or "thermophilic" bacteria (see Figure 63). Some raw materials, like algae, require this higher range for digestion. But digesters are not commonly operated at this higher range because: (1) most materials digest well at the lower range, (2) the thermophilic bacteria

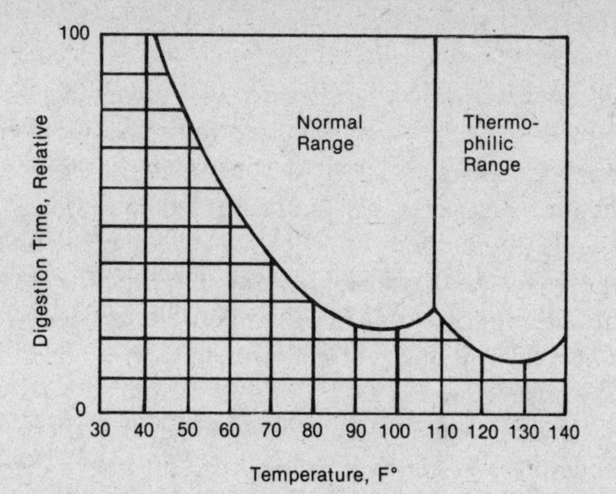

FIGURE 63. Digestion time and temperature.

are *very* sensitive to any changes in the digester, (3) the sludge they produce is of poor fertilizer quality, and (4) because it is difficult to maintain such a high temperature, especially in temperate climates.

The bacteria that produce methane in the "normal range"—90° to 95°F—are more stable and produce a high-quality sludge. It is not difficult to maintain a digester temperature of 95°F (see **Digester Heating**).

The same mass of manure will digest twice as fast at 95° than it will at 60° (see Figure 63) and it produces nearly 15 times more gas in the same amount of time! In Figure 64 it can be seen how a different amount of gas is produced when the digester is kept at 60°F than when it is kept at 95°F.

TABLE 14. Problems with pH

CONDITION	POSSIBLE REASONS	"CURE"
Too acid (pH 6 or less)	1) Adding raw materials too fast	Reduce feeding rate; Add a little ammonia
	2) Wide temperature fluctuation	Stabilize temperature
	3) Toxic substances	
	4) Build-up of scum	Remove scum
Too Alkaline (pH 9 or more)	1) Initial raw material too alkaline	Patience
		Never put acid into digester

Raw Materials

The amount and characteristics of organic materials (both plant and animal wastes) available for digestion vary widely. In rural areas, the digestible material will depend upon the climate, the type of agriculture practiced, the animals used and their degree of confinement, the methods of collecting wastes, etc. There are also degrees of quality and availability unique to urban wastes. Because of all these things, it is practically impossible to devise or use any formula or rule-of-thumb method for determining the amount and quality of organic wastes to be expected from any given source. There is, however, some basic information which is useful when you start wondering how much waste you can feed your digester.

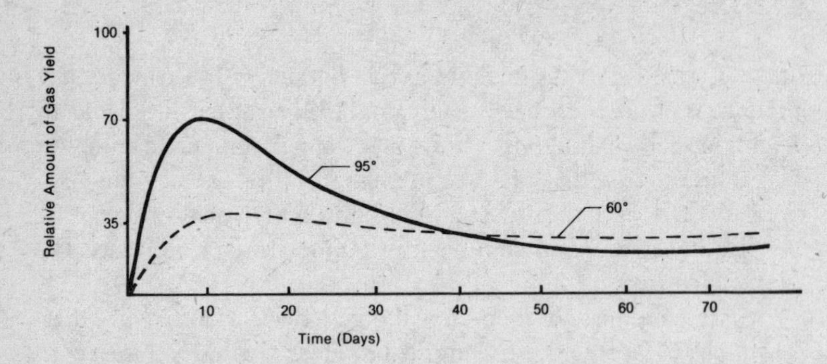

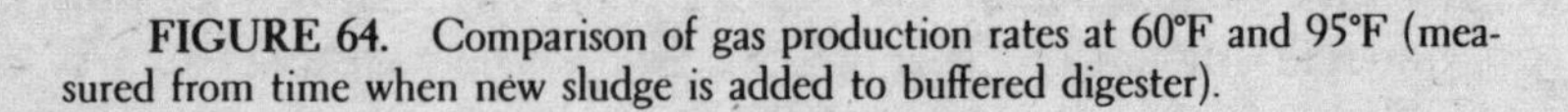

FIGURE 64. Comparison of gas production rates at 60°F and 95°F (measured from time when new sludge is added to buffered digester).

DIGESTIBLE PROPERTIES OF ORGANIC MATTER

When raw materials are digested in a container, only part of the waste is actually converted into methane and sludge. Some of it is indigestible to varying degrees, and it accumulates in the digester or passes out with the effluent and scum. The "digestibility" and other basic properties of organic matter are usually expressed in the following terms[15]:

Moisture: The weight of water lost upon drying at 220°F until no more weight is lost.

Total Solids (TS): The weight of dry material remaining after drying as above. TS weight is usually equivalent to "dry weight." (However, if you dry your material in the sun, assume that it will still contain around 30 percent moisture.) TS is composed of digestible organic or "Volatile Solids", and indigestible residues or "Fixed Solids."

Volatile Solids (VS): The weight of organic solids burned off when dry material is "ignited" (heated to around 1000°F). This is a handy property of organic matter to know, since VS can be considered as the amount of solids actually converted by the bacteria.

Fixed Solids (FS): Weight remaining after ignition. This is biologically inert material.

As an example, consider the make-up of fresh chicken manure.[16]

If we had 100 pounds of fresh chicken manure, 72 to 80 pounds of this would be water, and only 15 to 24 pounds (75 to 80 percent Volatile Solids of the 20 to 28 percent Total Solids) would be available for actual digestion (see Figure 65).

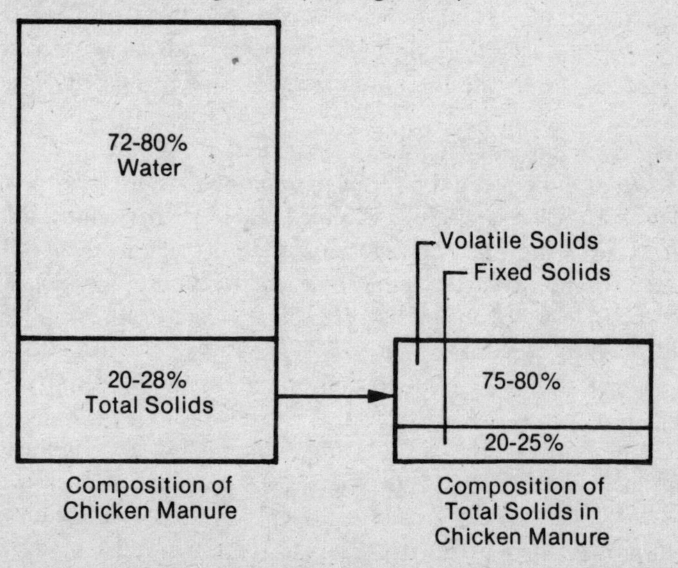

FIGURE 65. Properties of chicken manure.

AMOUNT OF MANURE COLLECTABLE

When you see a table which shows the amount of manure produced by different kinds of livestock, it's important to know that the amount on the table may not be the amount that is actually available from your animals. There are three major reasons for this:

1) The Size (Age) Of The Animal

Consider the total wet manure production of different sized pigs:

TABLE 15.

HOG WEIGHT	TOTAL MANURE POUNDS/DAY	FECES	URINE	RATIO MANURE/ HOG WEIGHT
40–80	5.6	2.7	2.9	1:11
80–120	11.5	5.4	6.1	1:9
120–160	14.6	6.5	8.1	1:10
160–200	17.6	8.5	9.1	1:10

So the size (age) of your livestock has a lot to do with the amount of manure produced. Notice that the ratio of total manure production to the weight of the pig is fairly constant. It is likely that similar ratios could be worked out for other kinds of livestock, enabling you to estimate the production of manure from the size of livestock.

2) The Degree of Livestock Confinement

Often the values given for manure production are for commercial animals which are totally confined. All of their manure can be collected. On the homestead or small farm, total confinement of the livestock is not always possible or even desirable. (Foraging and uncrowded livestock are less likely to contact diseases and more likely to increase the quality of their diet with naturally occurring foods.) Because of this, a large proportion of the manure is deposited in fields, and is thus hard to collect.

For example, the fresh manure production of commerical chickens in total confinement is about 0.4 pounds per chicken per day.[18, 19] However, for small-scale operations like homesteads and small farms, where preference tends to favor the well-being of the chickens rather than the economics of egg production, chickens are often allowed to forage all day and confined only at night. In such cases, only manure dropped during the night from roosts can be conveniently collected. In our experience, this may amount to only about 0.1 to 0.2 pounds of fresh manure per day per adult chicken. Similar reasoning holds for other livestock.

3) The Kind of Manure that is Collected
   a) All the fresh excrement (feces and urine).
   b) All the fresh excrement plus the bedding material.
   c) Wet feces only.
   d) Dry feces only.

Manure Production and the Livestock Unit

Keeping in mind all these factors that can affect the type and amount of manure that can be collected, we can assemble a general manure production table. The table only shows rough average values obtained from many sources. Values are expressed as the amount in pounds of wet manure, dry manure, and volatile solids that could be expected from various adult livestock per year. For the table, an adult animal is: cow—1000 pounds; horse—850 pounds; swine—160 pounds; human—150 pounds; sheep—67 pounds; turkey—15 pounds; duck—6 pounds; chicken—3½ pounds.

Table 16 enables us to get some idea of the production of readily digestible material (volatile solids) from different animals. Only the feces is considered for cows, horses, swine, and sheep, since their urine is difficult to collect. However, for humans and fowl, both urine and feces are given, since they are conveniently collected together.

The relative values of digestible wastes produced are not given in pounds of manure per animal per day, but in a more convenient relative unit called the "Livestock Unit." The table shows that on the average one medium horse would produce as much digestible manure as four large pigs, 12½ ewes, 20 adult humans, or 100 chickens.

Carbon to Nitrogen Ratio (C/N)

From a biological point of view, digesters can be considered as a culture of bacteria feeding upon and converting organic wastes.

The elements carbon (in the form of carbohydrates) and nitrogen (as protein, nitrates, ammonia, etc.) are the chief foods of anaerobic bacteria. Carbon is utilized for energy and the nitrogen for building cell structures. These bacteria use up carbon about 30 times faster than they use nitrogen.

Anaerobic digestion proceeds best when raw material fed to the bacteria contains a certain amount of carbon and nitrogen together. The carbon to nitrogen ratio (C/N) represents the proportion of the two elements. A material with 15 times more carbon than nitrogen would have a C/N ratio of 15 to 1 (written C/N = 15/1, or simply 15).

A C/N ratio of 30 (C/N = 30/1, 30 times as much carbon as nitrogen) will permit digestion to proceed at an optimum rate, if other conditions are favorable, of course. If there is too much carbon (high C/N ratio; 60/1 for example) in the raw wastes, nitrogen will be used up first, with carbon left over. This will make the digester slow down. On the other hand, if there is too much nitrogen (low C/N ratio; 30/15 for example, or simply 2), the carbon soon becomes exhausted and fermentation stops. The remaining nitrogen will be lost as ammonia gas (NH_3). This loss of nitrogen decreases the fertility of the effluent sludge.

There are many standard tables listing the C/N ratios of various organic materials, but they can be very misleading for at least two reasons:

1) The ratio of carbon to nitrogen measured chemically in the laboratory is often not the same as the ratio of carbon and nitrogen available to the bacteria as food (some of the food could be indigestible to the bacteria; straw, lignin, etc.).
2) The nitrogen and carbon content of even a specific kind of plant or animal waste can vary tremendously according to the age and growing conditions of the plant; and the diet, age, degree of confinement, etc., of the animal.

Nitrogen: Because nitrogen exists in so many chemical forms in nature (ammonia, NH_3; nitrates, NO_3; proteins, etc.), there are no reliable "quick" tests for measuring the total amount of nitrogen in a given material. One kind of test might measure the organic and ammonia nitrogen (the Kjeldahl test), another might measure the nitrate/nitrite nitrogen, etc. Also, nitrogen can be measured in

Table 16. Manure and the Livestock Unit

Average Adult Animal	Lbs/Day/Animal Urine	Lbs/Day/Animal Feces	Total Solids/Day 20% of Feces	Volatile Solids/Day 80% of TS– 85% for Swine	Livestock Units
BOVINE	20	52	10	8.0	
Bulls					130–150
Dairy cow					120
Under 2 yrs					50
Calves					10
HORSES	8	36	7	5.5	
Heavy					130–150
Medium					100
Pony					50–70
SWINE	4.0	7.5	1.5	1.3	
Boar, sow					25
Pig >160 lbs					20
Pig <160 lbs					10
Weaners					2
SHEEP	1.5	3	0.5	0.4	
Ewes, rams					8
Lambs					4

Portion	Amount	%TS	TS/Day	%VS	VS/Day	Livestock Units
HUMANS Urine	2 pints, 2.2 lbs	6%	.13	75%	.10	5
Feces	0.5 lbs	27%	.14	92%	.13	
Both	2.7 lbs	11%	.3	84%	.25	
FOWL Geese, turkey	0.5					2
Ducks						1.5
Layer chicken	0.3	35%	.1	65%	.06	1
Broiler chicken	0.1					

terms of wet weight, dry weight, or volatile solids content of the material; all of which will give different values for the proportion of nitrogen. Finally, the nitrogen content of a specific kind of manure or plant waste can vary, depending on the growing conditions, age, diet, and so forth.

For example, one study reported a field of barley which contained 39 percent protein on the 21st day of growth, 12 percent protein on the 49th day (bloom stage), and only 4 percent protein on the 86th day.[20] You can see how much the protein nitrogen depends on the age of the plant.

The nitrogen content of manure also varies a great deal. Generally, manures consist of feces, urine, and any bedding material (straw, corn stalks, hay, etc.) that may be used in the livestock shelters. Because urine is the animal's way of getting rid of excess nitrogen, the nitrogen content of manures is strongly affected by how much urine is collected with the feces.

For example, birds naturally excrete feces and urine in the same load, so that the nitrogen content of chickens, turkeys, ducks, and pigeons are highest of the animal manures in nitrogen content. Next in nitrogen content, because of their varied diets or grazing habits, are humans, pigs, sheep, and then horses. Cattle and other ruminants (cud chewers) which rely on bacteria in their gut to digest plant foods, have a low content of manure nitrogen because much of the available nitrogen is used to feed their intestinal bacteria. (See Figure 66.)

Even with the same kind of animal there are big differences in the amount of manure nitrogen. For example, stable manure of horses may have more nitrogen than pasture manure because feces and urine are excreted and collected in the same small place.

Since there are so many variables, and because anaerobic bacteria can use most forms of nitrogen, the *available nitrogen* content of organic materials can best be generalized and presented as *total nitrogen* (percent of dry weight).

Carbon: Unlike nitrogen, carbon exists in many forms which are not directly useable by bacteria. The most common indigestible form of carbon is lignin, a complex plant compound which makes land plants rigid and decay-resistant. Lignin can enter a digester either directly with plant wastes themselves or indirectly as bedding or undigested plant food in manure. Thus, a more accurate picture of the C part of the C/N ratio is obtained when we consider the "non-lignin" carbon content of plant wastes.

CALCULATING C/N RATIOS

Table 17 can be used to calculate roughly the C/N ratios of mixed raw materials. Consider the following examples:

Example 1: Calculate the C/N ratio of 50 lbs horse manure (C/N =25) plus 50 lbs dry wheat straw (C/N=150).
Nitrogen in 50 lbs horse manure = 2.3% × 50 = 1.2 lbs
Carbon in 50 lbs horse manure = 25 times more than nitrogen = 25 × 1.2 = 30 lbs
Nitrogen in 50 lbs wheat straw = 0.5% × 50 = .25 lbs
Carbon in 50 lbs wheat straw = 150 times more than nitrogen = 150 × .25 = 37.5 lbs

	MANURE	STRAW	TOTAL
Carbon	30	37.5	67.5 lbs
Nitrogen	1.2	.25	1.45 lbs

C/N ratio = 67.5/1.45 = 46.5

Although a bit high, this would be a satisfactory ratio for most digestion purposes.

Example 2: Calculate the C/N ratio of 8 lbs grass clippings (C/N =12) and 2 lbs of chicken manure (C/N=15).
Nitrogen in 8 lbs grass clippings = 4% × 8 = .32 lbs
Carbon in 8 lbs grass clippings = 12 times more than nitrogen = 3.8 lbs
Nitrogen in 2 lbs chicken manure = 6.3% × 2 = .13 lbs
Carbon in 2 lbs chicken manure = 15 times more than nitrogen = 1.9 lbs

	MANURE	GRASS	TOTAL
Carbon	3.8	1.9	5.7
Nitrogen	.32	.13	.45
C/N ratio = 5.7/.45 = 12.6			

The C/N ratio of this mixture is low. We might want to add a higher proportion of chicken manure since it contains more carbon per weight than the grass.

The following table is a summary of the important chemical properties of organic materials. Values are averages derived from many sources and should be used only for approximation.

TABLE 17. Carbon and Nitrogen Values of Wastes

	TOTAL NITROGEN % DRY WEIGHT	C/N RATIO
ANIMAL WASTES		
Urine	16	0.8
Blood	12	3.5
Bone meal		3.5
Animal tankage		4.1*
Dry fish scraps		5.1*
MANURE		
Human, feces	6	6–10
Human, urine	18	
Chicken	6.3	15
Sheep	3.8	
Pig	3.8	
Horse	2.3	25*
Cow	1.7	18*
SLUDGE		
Milorganite		5.4*
Activated	5	6
Fresh sewage		11*
PLANT MEALS		
Soybean		5
Cottonseed		5*
Peanut hull		36*
PLANT WASTES		
Hay, young grass	4	12
Hay, alfalfa	2.8	17*
Hay, blue grass	2.5	19
Seaweed	1.9	19
Non-legume vegetables	2.5–4	11–19
Red clover	1.8	27
Straw, oat	1.1	48
Straw, wheat	0.5	150
Sawdust	0.1	200–500

Nitrogen is total nitrogen dry weight and carbon is either total carbon (dry weight) or (*) non-lignin carbon (dry weight).

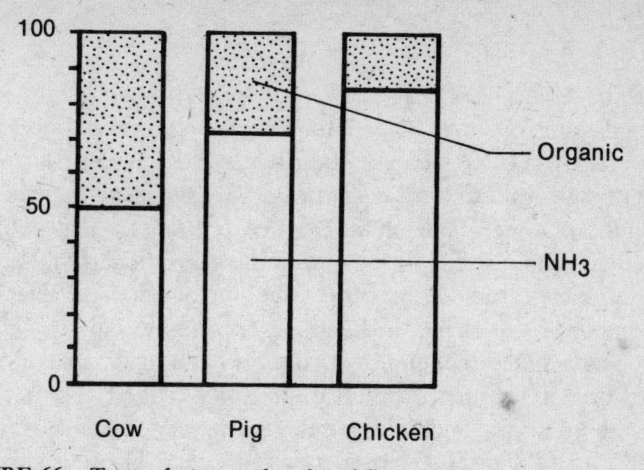

FIGURE 66. Types of nitrogen found in different kinds of manure.

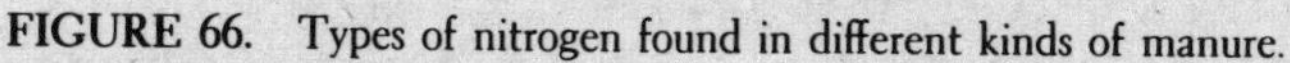

The Gas

The gas produced by digestion, known as marsh gas, sewage gas, dungas, or bio-gas, is about 70 percent methane (CH_4) and 29 percent carbon dioxide (CO_2) with insignificant traces of oxygen and sulfurated hydrogen (H_2S) which gives the gas a distinct odor. (Although it smells like rotten eggs, this odor has the advantage of being able to trace leaks easily.)

The basic gas-producing reaction in the digester is: carbon plus water equals methane plus carbon dioxide ($2C + 2H_2O = CH_4 + CO_2$). The methane has a specific gravity of .55 in relation to air. In other words, it is about half the weight of air and so rises when released to the atmosphere. Carbon dioxide is more than twice the weight of air, so the resultant combination of gases, or simply bio-gas, when released to atmosphere, will rise slowly and dissipate.

TABLE 18. General Composition of Bio-Gas Produced From Farm Wastes (from many references)

CH_4	methane	54 –70%
CO_2	carbon dioxide	27 –45%
N_2	nitrogen	0.5– 3%
H_2	hydrogen	1 –10%
CO	carbon monoxide	0.1%
O_2	oxygen	0.1%
H_2S	hydrogen sulfide	trace

FUEL VALUE

The fuel value of bio-gas is directly proportional to the amount of methane it contains (the more methane, the more combustible the bio-gas). This is because the gases, other than methane, are either non-combustible, or occur in quantities so small that they are insignificant. Since tables of "Fuel Values of Bio-Gas" may not show how much combustible methane is in the gas, different tables show a wide variety of fuel values for the same kind of gas, depending upon the amount of methane in the gas of each individual table.

As a general rule, pure methane gas has a heat value of about 1,000 British Thermal Units (Btu) per cubic foot (ft^3). One Btu is the amount of heat required to raise one pound (one pint) of water by 1°F. Five ft^3, or 5000 Btu of gas is enough to bring ½ gallon of water to boiling and keep it there 20 minutes. If you have a volume of bio-gas which is 60 percent methane, it will have a fuel value of about 600 Btu/ft^3, etc.

TABLE 19. Fuel Value of Bio-Gas and Other Major Fuel Gases

FUEL GAS	FUEL VALUE (BTU/FT^3)
Coal (town) gas	450–500
Bio-gas	540–700
Methane	896–1069
Natural gas (methane or propane-based)	1050–2200
Propane	2200–2600
Butane	2900–3400

The composition and fuel value of bio-gas from different kinds of organic wastes depends on several things:

1) The temperature at which digestion takes place.
2) The nature of the raw material. According to Ram Bux Singh:[21] "pound for pound, vegetable waste results in the production of seven times more gas than animal waste." In our experience, pressed plant fluids from succulent plants (cactus), greatly increases the amount of gas produced, but certainly not by a factor of seven. Harold Bates (who built the famous chicken manure car) has noted that more gas is produced from manure with a little straw added. But we

are more interested in the production of methane than bio-gas. Laboratory experiments[22, 23] have shown that plant materials produce bio-gas with a high proportion of carbon dioxide. So, the extra gas produced by plants may be less valuable for our purposes of fuel production.

The general quality of bio-gas can be estimated from the C/N ratio of the raw materials used. (See Table 20.)

TABLE 20. Gas Production According to C/N Ratios of Raw Wastes

		METHANE	CO_2	HYDROGEN	NITROGEN
C/N Low (high nitrogen)	blood, urine	little	much	little	much
C/N High (low nitrogen)	sawdust, straw, sugar and starches such as potatoes, corn, sugar beet wastes	little	much	much	little
C/N Balanced	manures, garbage	much	some	little	little

With good temperature and raw materials, 50 to 70 percent of the raw materials fed into the digester will be converted to bio-gas.

Amount of Gas From Different Wastes

The actual amount of gas produced from different raw materials is extremely variable, depending upon the properties of the raw material, the temperature, the amount of material added regularly, etc. Again, for general rule-of-thumb purposes, the following combinations of wastes from a laboratory experiment can be considered as minimum values: (Table 21)

Other values for gas production are from working digester operations. These are shown as cubic feet of gas produced by the total solids and are more liberal values than in Table 21.

As an example, suppose we had 100 chickens which were allowed to forage during the day, but were cooped up at night, so that only about half of their manure was collectable. At 0.1 lb/chicken/

TABLE 21. Cubic Feet of Gas Produced by Volatile Solids of Combined Wastes[24]

MATERIAL	PROPORTION	FT^3 GAS PER LB VS ADDED	CH_4 CONTENT OF GAS	FT^3 OF CH_4 PER LB VS ADDED
Chicken manure	100%	5.0	59.8	3.0
Chicken manure & paper pulp	31% 69%	7.8	60.0	4.7
Chicken manure & newspaper	50% 50%	4.1	66.1	2.7
Chicken manure & grass clippings	50% 50%	5.9	68.1	4.0
Steer manure	100%	1.4	65.2	0.9
Steer manure & grass clippings	50% 50%	4.3	51.1	2.2

day this would amount to about 10 lbs of wet or 3.5 lbs dry (see Table 16) manure per day. Other conditions being equal, this could be equivalent to about 20 to 40 ft³ of bio-gas (assuming 60 percent methane) 12 to 24 ft³ of methane gas per day.

TABLE 22. Bio-Gas Produced By Total Solids of Wastes

MANURE	FT^3/LB OF DRY MATTER (TS)
Pig	6.0– 8.0
Cow (India)	3.1– 4.7
Chicken	6.0–13.2
Conventional sewage	6.0– 9.0

Digesters

BASIC DIGESTER DESIGN

Digesters can be designed for batch-feeding or for continuous feeding. With batch digesters a full charge of raw material is placed into the digester which is then sealed off and left to ferment as long as gas is produced. When gas production has ceased, the digester is emptied and refilled with a new batch of raw materials.

Batch digesters have advantages where the availability of raw materials is sporadic or limited to coarse plant wastes (which contain undigestible materials that can be conveniently removed when batch digesters are reloaded). Also, batch digesters require little daily attention. Batch digesters have disadvantages, however, in that a great deal of energy is required to empty and load them; also gas and sludge production tend to be quite sporadic. You can get around this problem by constructing multiple batch digesters connected to the same gas storage. In this way individual digesters can be refilled in staggered sequence to ensure a relatively constant supply of gas. Most early digesters were of the batch type.

With continuous-load digesters, a small quantity of raw material is added to the digester every day or so. In this way the rate of production of both gas and sludge is more or less continuous and reliable. Continuous-load digesters are especially efficient when raw materials consist of a regular supply of easily digestible wastes from nearby sources such as livestock manures, seaweed, river or lake flotsam, or algae from production sludge-ponds. The first continuous-load digester seems to have been built in India by Patel in 1950.[25]

Continuous-feeding digesters can be of two basic designs: vertical-mixing or displacement (see Figure 67). Vertical-mixing digesters consist of vertical chambers into which raw materials are added. The slurry rises through the digester and overflows at the top. In

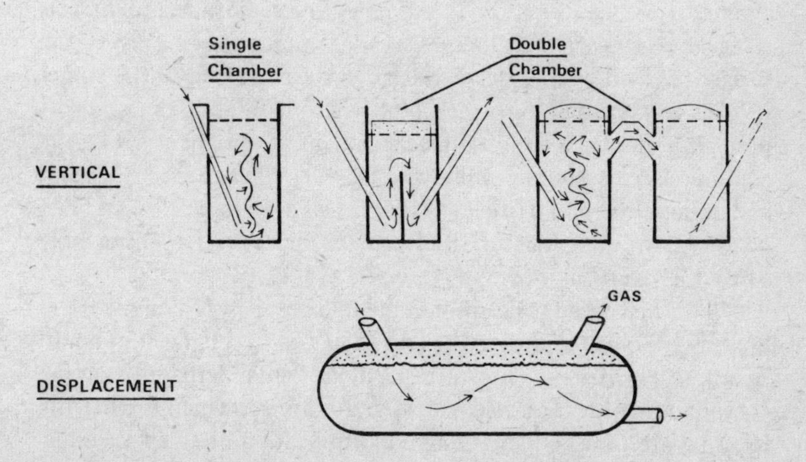

FIGURE 67. Types of continuous-feeding digesters.

single-chamber designs the digested or "spent" slurry can be withdrawn directly from effluent pipes. In double-chamber designs the spent slurry, as it overflows the top, flows into a second chamber where digestion continues to a greater degree of completion.

Displacement digesters consist of a long cylinder lying parallel to the ground (e.g., inner tubes, oil drums welded end on end, tank cars, etc.). As it is digested the slurry is gradually displaced toward the opposite end, passing a point of maximum fermentation on the way.

The displacement digester design seems to have distinct advantages over vertical-mixing designs popularized in India: (1) In vertical-mixing digesters raw material is subject to a vertical pumping motion and often escapes the localized action of digesting bacteria. Slurry introduced at one time can easily be withdrawn soon afterwards as incompletely digested material. In displacement digesters slurry must pass an area of maximum fermentation activity so that all raw materials are effectively digested (much like the intestines of an animal). (2) From a practical point of view, displacement digesters are easier to operate. If digester contents begin to sour for one reason or another, strongly buffered material at the far end can be recirculated efficiently by simply reversing the flow of material along the line of the cylinder. In addition, raw materials can be digested to any desired degree without the need for constructing additional chambers or digesters. (3) The problem of scum accumulation is reduced in displacement digesters. Since scum forms evenly on the surface of the digesting slurry, the larger the surface area, the longer it takes to accumulate to the point where it inhibits digestion. A prone cylinder has a larger surface area than an upright one. (4) Any continuous-load digester will eventually accumulate enough scum and undigested solid particles so that it will have to be cleaned. The periodical washing out of displacement digesters is considerably easier than vertical-mixing digesters.

The first large-scale displacement digester was designed and built by L. John Fry during the late 1950's on his pig farm in South Africa.[26, 27] Mr. Fry, now a resident of Santa Barbara, is acting consultant for the New Alchemy digester project which is currently focusing attention on the design and utilization of small-scale displacement digesters.

RAW MATERIALS AND DIGESTER DESIGN

Plant Wastes: The primary advantage to plant wastes is their availability. Their disadvantage for a small farm operation is that plant wastes can often be put to better use as livestock feed or compost. Also, plants tend to be bulky and to accumulate lignin and other indigestible materials that must be regularly removed from digesters. This severely limits the use of plant wastes in continuous-feeding digesters.

There seem to be three possible ways to take advantage of plant wastes in continuous digesters: (1) Press plant fluids out of succulent plants (e.g., cacti, iceplant, etc.) and digest juices directly, or use them as a diluter for swill. (2) Culture algae for digestion. (3) Digest plants not containing lignin (e.g., seaweed).

Animal Manures: The main advantage to animal manures, with respect to continuous digesters, is that they are easy to collect (with proper design of livestock shelters) and easy to mix as slurry and load into digesters. Successful continuous digesters have been set up using pig manure,[28–30] cow dung[31–33] and chicken manure.[34] The consensus seems to be that, among animal manures, chicken manure "is easily digested, produces large quantities of gas, and makes a fertilizer very high in nitrogen."[35]

Human Waste: Human waste or "night soil" has long been used as a fertilizer, especially in the Orient.[36, 37] However, there seems to be little information on using human wastes as raw materials for anaerobic digesters. A few ideas involving outhouses and latrines are described by Gotaas in his chapter, *Manure and Night Soil Digesters for Methane Recovery on Farms and in Villages.*[38] It seems possible, also, that digesters could be incorporated into aerobic dry toilet designs of the "Clivus" type.[39] This may be especially fruitful since the main drawback to using human wastes from flush toilets is the excess water that is carried with it which inhibits digestion. A well-designed privy digester which paid special attention to the transmission of diseases peculiar to humans would be a real asset to homestead technology. A solution to this problem would be welcomed. One suggestion is a seat with a clip-on plastic bag. When filled it could be dropped into a digester intact. The plastic would have to be a material which would decompose only in the presence of methane bacteria, or liquids generally after so many hours.

LOADING RATE, DETENTION TIME, AND DIGESTER SIZE

In calculating the size of a continuous-load digester the most important factors are loading rate and detention time.

Loading Rate: Is defined as the amount of raw material (usually pounds of volatile solids) fed to the digester per day per cubic foot of digester space. Most municipal sewage plants operate at a loading rate of .06–.15 lb VS/day/ft³. With good conditions, much higher rates are possible (up to .4 lbs VS/day/ft³). Again, as with most aspects of digesters, the optimum situation is a compromise. If you load too much raw material into the digester at a time, acids will accumulate and fermentation will stop. The main advantage to a higher loading rate is that by stuffing a lot into a little space, the size (and therefore cost) of the digester can be reduced.

Example: Suppose you had 10 lbs of fresh chicken manure (total manure from about 30 chickens) available for digestion every day: 10 lbs fresh chicken manure = 2.3 volatile solids (Table 4). At a loading rate of .2 lbs VS/day/ft³ this would require a digester 2.3/.2 = 12 ft³ in volume (about the size of two 50-gallon drums). At a loading rate of .1 lb VS/day/ft³, this would double the necessary size of the digester with the same amount of manure.

Detention Time: This is the number of days that a given mass of raw material remains in a digester. Since it is very difficult to load straight manure into a digester it is usually necessary to dilute it with water into a slurry. If too much water is added, the mixture will become physically unstable and settle quickly into separate layers within the digester, thus inhibiting good fermentation. The general rule-of-thumb is a slurry about the consistency of cream. The important point here is that as you dilute the raw material you reduce its detention time.

Example: The volume of 10 lbs of fresh chicken manure is about .2 ft³. If this is diluted 1:1 with water the volume becomes about .4 ft³. With the 12 ft³ digester described above, this would mean a detention period of 12/.4 = 36 days. If the manure were diluted more, say 2:1, the volume would be .6 ft^3 and the detention period would be reduced to $12/.6 = 20$ days. Up to a point, then (usually no less than 6 percent solids), diluting raw materials will produce the same amount of gas in a shorter period of time.

These relationships between loading rate, detention time, and digester size reveal themselves more clearly after direct experience with continuous-load digesters. However, generalities can be of some use in the beginning.

Heating Digesters

For the most efficient operation, especially in temperate climates, digesters should be supplied with an external supply of heat to keep them around 95°F; there are several ways to do this. Methods which heat the outside of digesters (e.g., compost piles, light bulbs, and water jackets) could be more effectively used as insulation since much of their heat dissipates to the surroundings. (Since digesters should be constantly warmed rather than sporadically heated, compost "blankets" are not very practical unless you coordinate a regular program of composting with digestion.) Similarly, green houses built over digesters tend to overheat the digester during the day and cool it down at night.

The most effective method of keeping digesters warm is to circulate heated water through pipes or coils placed *within* the digester. The water can be heated by solar collectors or by water boilers heated with methane.

Gas-heated water boilers are a good idea since they allow the digestion process to feed back on itself, thus increasing efficiency. One practical gas-heater design we have used is shown in Figure 60. The thermostat in the water boiler is set at 140°F because slurry will cake on surfaces (e.g., the water coils) warmer than this. The digester thermostat is set at the optimum 95°F. Until the digester begins producing methane, propane can be used as a fuel source for the water boiler.

For optimum heat exchange within the digester, a ratio of 1 ft^2 coil area per 100 ft^3 of digester volume is recommended.[40]

INSULATING DIGESTERS

A word of caution if you insulate your digester. Methane is not only combustible but highly explosive when it makes up more than 9 percent of the surrounding air in confined spaces. If you use synthetic insulation, avoid porous materials such as spun glass which can trap gas mixtures. It's easy to scrounge styrofoam sheets since they are so commonly used as packing material and regularly discarded. Styrofoam is one of the best insulating materials, although it is slightly flammable.

Using Gas

PROPERTIES OF METHANE

Specific Gravity (air = 1.0)	.55
Dry Weight, lb/ft³	.04 (gas)
Liquid Weight, lb/gal	3.5 (liquid)
Fuel Value, Btu/ft³	950–1050
Air for Combustion, ft³/ft³	9.5
Flammability in Air, % Methane	5–14

USES OF METHANE

General: Methane can of course be used in any appliance or utility that uses natural gas. The natural gas requirements of an average person with a U.S. standard of living is about 60 ft^3/day. This is equivalent to 10 pounds of chicken or pig manure per day (seven pigs and 100 chickens) or 20 pounds of horse manure (about two horses). Other uses and methane requirements are listed in Table 23.

Heat Engines: Methane, the lightest organic gas, has two fundamental drawbacks to its use in heat engines: it has a relatively low fuel value (Table 19), and it takes nearly 5,000 psi to liquefy it for easy storage. (87.7 ft^3 methane gas = 1 gallon of liquid methane or 1 ft^3 methane gas = 9 tablespoons liquid methane.) So a great deal of storage is required of methane for a given amount of work. For

TABLE 23. Uses for Methane

USE	FT^3	RATE
Lighting	2.5	per mantle per hour
Cooking	8–16	per hour per 2–4 in burner
	12–15	per person per day
Incubator	.5–.7	ft^3 per hour per ft^3 incubator
Gas Refrigerator	1.2	ft^3 per hour per ft^3 refrigerator
Gasoline Engine*		
CH_4	11	per brake horse-power per hour
Bio-Gas	16	per brake horse-power per hour
For Gasoline		
CH_4	135–160	per gallon
Bio-Gas	180–250	per gallon
For Diesel Oil		
CH_4	150–188	per gallon
Bio-Gas	200–278	per gallon

*25% efficiency

comparison, propane liquefies around 250 psi. Consider the following example where methane is compressed to just 1,000 psi in a small bottle and used to power a rototiller of 6 brake horsepower:

<u>Example:</u>

1 hph = 2540 Btu = effective work

Fuel value of methane = 950 Btu/ft^3

TV (tank volume) = 2-ft-by-6-in–cylinder = 678 in^3 = 0.39 ft^3

TP (tank pressure) = 1000 psi = 68 atmos

EV (effective volume) = (TP) (TV) = 26.7 ft^3 = 25,300 Btu

hp = brake horsepower of engine

hr = hours of running

x = heat value of gas (Btu/ft^3)

y = efficiency of engine (25% for conventional gas engines

<u>Methane Gas Consumption (G) (ft^3)</u> of general heat engine:

$$G = \frac{(hp)\,(2540)\,(hr)}{(x)\;(y)}$$

of gasoline engine on methane:

$$G = \frac{(hp)\,(2540)\,(hr)}{(0.25)\,(9.50)} = (10.7\ ft^3/hph)$$

for a 6 brake hp rototiller

$$G = (hp)\,(hr)\,(10.7\ ft^3/hph) = (64.2\ ft^3/hr)(hr)$$

Operating Time (OT)

$$OT = \frac{EV}{G} = 0.414 \text{ hr} = 25 \text{ minutes/tank}$$

Useful
Work = 2.5 hph = 6,350 Btu =
Supplied

(25,300 Btu/tank) (25% efficiency)

At 25 percent compressor efficiency it would take .52 hph to compress the gas (1320 Btu). In other words, it would take 1320 Btu to compress 25,300 Btu worth of gas that provides 6,350 Btu worth of work. Clearly the system is not very "efficient" in the sense that 21 percent of the *resulting* work energy is needed for compression while 75 percent of the *available* energy is lost as heat.

Methane has been used in tractors[41, 42] and automobiles.[43] The gas bottles carried by such vehicles are often about 5 feet long by 9 inches diameter (1.9 ft^3) charged to 2800 psi so that about 420 ft^3 of methane is carried (about 3½ gallons gasoline). However, it seems that the most efficient use of methane would be in stationary heat engines located near the digester (e.g., compressors and generators). There are two reasons for this: (1) The engine's waste heat can be recirculated in digester coils instead of dissipating in the open. (2) Gas can be used directly as it is produced, without the need of compressors.

For example, bio-gas produced from pig manure was used at ordinary pressures by John Fry to power a Crossley Diesel engine. The diesel ran an electric generator and the waste heat was recirculated directly back into the digesters.[44] It is likely that bio-gas produced from mixed wastes would have to be "scrubbed" of corrosive hydrogen sulfide (by passing through iron filings), and possibly CO_2 (by passing through lime water).

EFFICIENCY OF DIGESTION

The efficiency of anaerobic digestion can be estimated by comparing the energy available in a specific amount of raw material to the energy of the methane produced from that material. Four such estimates are given below (Figure 68).

Given the heat value of raw materials and their resultant gas, it seems fair to conclude that anaerobic digestion is about 60–70 percent "efficient" in converting organic waste to methane. However, it would probably be more accurate to call this a conversion rate since, like all biological processes, a great deal of energy is required to maintain the system, and most of this extra energy is not included in the conversion. For example, consider how much energy is needed just to keep a digester warm in a general temperate climate.

> Example: Direct-heating hot water boilers have an efficiency of about 70 percent. Gas engines have a power efficiency of 20–25 percent, and a water heating efficiency of about 50 percent.[45] As hot water heaters, then, heat engines are about as efficient as water boilers. In either case about 20–30 percent of the gas energy derived from digestion must be put back into the system to heat digesters. Without even considering the energy needed to collect raw materials or load and clean the digesters, the conversion efficiency of digestion should be closer to 50 percent.

Using Sludge

Sludge as a Fertilizer

Most solids not converted into methane settle out in the digester as a liquid sludge. Although varying with the raw materials used and the conditions of digestion, this sludge contains many elements essential to plant life: nitrogen, phosphorus, potassium, plus small amounts of metallic salts (trace elements) indispensable for plant growth such as boron, calcium, copper, iron, magnesium, sulfur, zinc, etc.

Nitrogen is considered especially important because of its vital role in plant nutrition and growth. Digested sludge contains nitrogen mainly in the form of ammonium (NH_4), whereas nitrogen in aerobic organic wastes (activated sludge, compost) is mostly in oxidized forms (nitrates, nitrites). Increasing evidence suggests that for many land and water plants ammonium may be more valuable as a nitrogen source than oxidized nitrogen; in the soil it is much less apt to leach away and more apt to become fixed to exchange particles (clay and humus). Likewise, important water algae appear to be able to utilize ammonium easier than nitrates.[46] Generally speaking, this is a reversal from the earlier belief by fertilizer scientists that oxidized nitrogen always presented the most available form of nitrogen for plants. Because of these things, it has been suggested that liquid digested sludge produces an increase of nitrogen comparable with those of inorganic fertilizers in equivalent amounts.[47]

Most of the information showing the poor fertilizer value of sludge has been based on municipal sewage sludge. It is a bad measure of the fertilizer value of digested sludge in general. (Municipal treatment flushes away all the fertilizer-rich liquid effluent.) In one case[48] digested sewage sludge was found to contain only about half the amount of nitrogen in fresh sewage, whereas elsewhere[49] digested pig manure was found to be 1.4 times richer in nitrogen content than raw pig manure. Similar results have been found with digested chicken manure.

Sludge from your digester can be recycled in a wide variety of ways, both on land and in water and pond cultures. The possibilities are many and only brief descriptions of potentials can be given here.

TABLE 24. Nitrogen Fertilizer Value of Various Sludges and Finished Compost (Taken from many sources)

	N (% DRY WT.)
Raw Sewage	1.0–3.5
Digested Sludge	
10 municipalities	1.8–3.1
12 Ohio municipalities	0.9–3.0
51 samples, 21 cities	1.8–2.3
General average	2.0
General average	1.0–4.0
Activated Sludge	
5 municipalities	4.3–6.4
General average	4.0–6.0
General average	4.0–7.0
Digested Manure Sludge	
Hog	6.1–9.1
Chicken	5.3–9.0
Cow	2.7–4.9
Finished Compost	
Municipal	.4–1.6
Garbage	.4–4.0
Garden	1.4–3.5

Sludge Gardening and Farming

The application of digested sludge to crops serves a double purpose since it is both a soil conditioner and fertilizer. The sludge humus, besides furnishing plant foods, benefits the soil by increasing the waterholding capacity and improving its structure. In some preliminary experiments with garden and house plants we have obtained astounding results with the use of sludge from our chicken manure digester. However, there are some things to consider first: (1) *Fresh* digested sludge, especially from manures, contains high amounts of ammonia, and in this state may act like a chemical fertilizer by force-feeding large amounts of nitrogen into the plant and increasing the accumulation of toxic nitrogen compounds.[50, 51] There is no direct evidence for this, but the possibility exists. For this reason it is probably best to let your sludge "age" for a few weeks in an open area (oil drums, plastic swimming pools, etc.), or in a closed container for a few months before using it on crops. The fresher it is the more you should dilute it with water before application. (2) The continued use of digested sludge in any one area tends to make soils acidic. You should probably add a little dolomite or limestone at regular intervals to your sludge plots, allowing at least two weeks interval between applications to avoid excess nitrogen loss. Unfortunately, limestone tends to evaporate ammonia so you may experience a temporary nitrogen loss when you apply it on your sludge plots. (3) Unlike digested municipal sludge, sludge from farm wastes does not contain large amounts of heavy metals or salts so there is little danger of applying it too heavily over a period of time. However, you should pay attention to the structure of your soil. If it contains a lot of clay, the sludge will tend to accumulate and possibly present problems in the root area of your plants. In general, keep close tabs on your sludge plots in the beginning until you become familiar with the behavior of sludge in your own particular soil.

Sludge-Pond Cultures

There are at least three general ways to integrate pond cultures with organic digesters: hydroponic crops, sludge-algae-fish and sludge-algae-methane systems. All have their advantages depending on local needs and resources.

Sludge Hydroponics: Hydroponics is the process of growing plants directly in nutrient solution rather than soil. The nutrients may consist of soluble salts (i.e., chemical fertilizers) or liquid organic wastes like digested sludge and effluent. Plants grown hydroponically in sludge-enriched solutions can serve a variety of purposes for organic digester operations: (1) They can do away with the cost and energy of transporting liquid fertilizer to crop lands since they can be grown conveniently near to digesters. (2) They tend to be more productive than conventional soil crops, and thus can serve as a high-yield source of fodder, compost, mulch, or silage. (3) They can serve as convenient high-yield sources of raw materials for the digester itself.

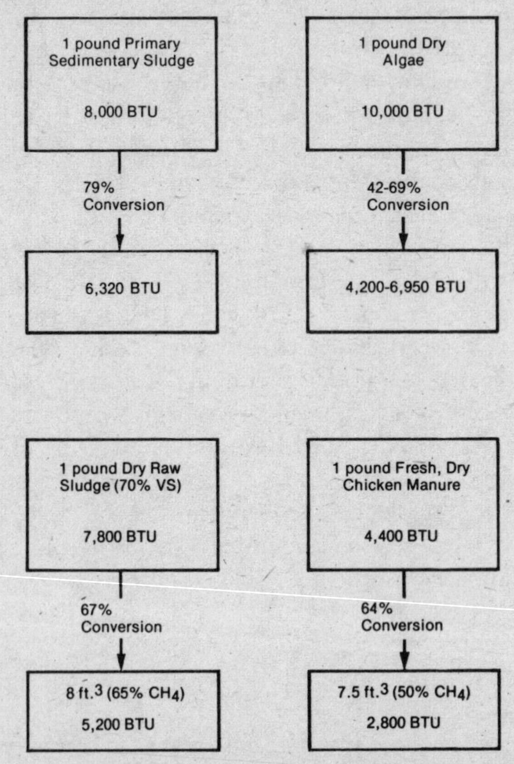

FIGURE 68. Efficiency of methane production from different materials.

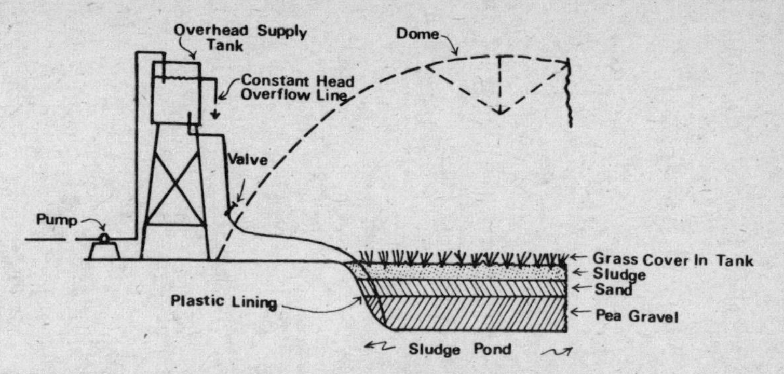

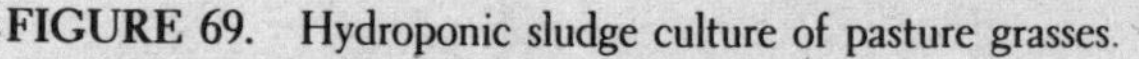
FIGURE 69. Hydroponic sludge culture of pasture grasses.

Information about the use of sludge to fertilize water plants comes from projects to treat waste water in run-off areas or "sewage lagoons."[52, 53] Some plants, for example water hyacinth, *Ipomoea repens,* and some cool season pasture grasses such as rye, fescue, and canary grass, have the ability to grow well in waste water and to take up great amounts of nutrients efficiently, thus helping to control polluted waters. These crops have the added advantage that they are easy to harvest for livestock feed, thus giving an efficient method of converting sludge nutrients into animal protein.

Usually, the plants are grown in shallow ponds filled with a diluted sludge solution. The process consists of slowly adding sludge under a gravel bed lining the pond and covered with a layer of fine sand. Over the sand, plants are sprouted in containers floating on the effluent that percolates up through the gravel and sand layers. After sprouting the grasses then root and anchor in the sand and gravel.

Sludge-Algae-Fish: The essence of the sludge-algae-fish or "aquaculture" system consists of placing sludge into ponds and stimulating the growth of algae. The algae are then used as feed for small invertebrates or fish growing in the pond. The idea is modeled after Oriental aquaculture systems.

During the last two years, under the direction of Bill McLarney, New Alchemy has established preliminary models for experimental fish cultures *(Tilapia).* A general description of small-scale fish farming methods using organic fertilizers and invertebrate fish food cultures has been presented elsewhere.[54–56]

Sludge-Algae-Methane: In the sludge-algae-methane system, green algae is grown in diluted sludge, then harvested, dried, and

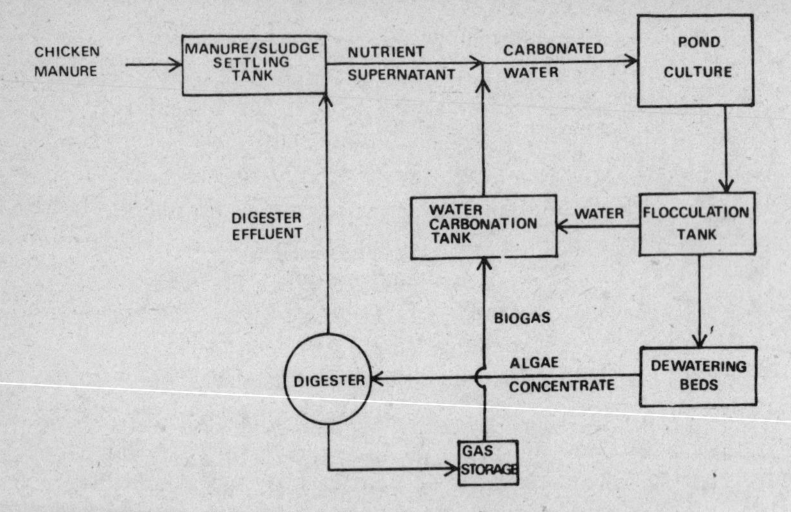

FIGURE 70. Flow diagram of sludge-algae-methane conversion system.

digested to produce methane for power and sludge for recycling. This procedure of transforming solar energy and sludge nutrients into the chemical energy of methane is potentially a very efficient and rapid biological process: (1) It is a closed nutritional system and (2) the rate of turnover is extremely high; organic matter is decomposed relatively quickly by anaerobic bacteria in the pond while it is most rapidly made by green algae. The complete sludge-algae-methane system involves a series of processes. The principle features of the system are integration of the algae culture with the gas in such a manner that nutrients and water are recycled from one process to the other (Figure 70). Most of the information concerning this system has been developed by researchers at the University of California at Berkeley in a manner that has real potential for the homestead or small farm.[58–61]

Our Four-Cow Bio-Gas Plant

Sharon and James Whitehurst

It was in May of 1972 that Ralph Hurd attended a meeting of Vermont's Natural Food and Farming Association, where he heard that Ram Bux Singh, acknowledged authority on producing methane gas from organic materials, was planning a visit to Vermont. We at Ral-Jim Farm are very interested in living a healthy life, and being of thrifty Yankee stock, we heat with wood, grow organic vegetables and put them up for winter, make our own whole wheat bread and gallons of yogurt! When we learned that farmers in India were producing gas from cow manure, we were immediately excited by an idea so much in keeping with our "homemade" mode of living. Ralph had heard of "swamp gas," and recalled an article some years back in *Organic Gardening and Farming* which had discussed methane. A little nosing around the state and several phone calls put us on Singh's waiting list of interested parties, and six weeks later, rather to our surprise, another phone call assured us that Mr. Singh was anxious to visit our farm and perhaps construct a plant there if we were interested. By the time Jim drove into the dooryard that afternoon with Singh in tow, a bio-gas plant at Ral-Jim Farm was a sure thing!

We should stress at this point that we are not engineers or chemists, nor do we know all there is to know about bio-gas. However, we did study intensively with Singh while he was with us, and we do have a plant which has been operating successfully for nearly two years, and in this time we have learned a lot about producing methane from animal wastes. We don't feel it is within the scope of this section to go into chemical technicalities, or to discuss the many modifications which can be made on the basic bio-gas plant structure. We think *Bio-Gas Plant* by Ram Bux Singh (see Bibliography) should be required reading for anyone seriously interested in the subject.

Working-class Indians have traditionally used dried "cow patties" for fuel. Although this manure provides sufficient heat for warmth and cooking when burned, there are two disadvantages to burning it. The manure produces quite an offensive, eye-watering, air-polluting smoke, and by using the manure in this manner, there is no residue with which to fertilize the land. In such a densely populated, underdeveloped country as India, robbing the land of much needed fertilizer can lead to serious food shortages.

Both these disadvantages are overcome, however, when the manure is fermented, rather than burned. There is no stinky smoke, and the bio-gas slurry that is removed from the digester and spread on the land boosts the nutrient-starved croplands to the point where yields per acre double and triple as the quality of the crops improves. In India, the methane gas is actually considered the secondary by-product, although in the U.S., where the majority of us have long depended upon commercial fertilizers, we would think of the energy (gas) as the prime reason for popularizing bio-gas plants.

Our plant is a very simple type. It has a 225-cubic foot capacity and it utilizes the manure from about four cows on a daily basis. It is a good-sized plant for a homesteader, enthusiastic organic gardener, or just anyone who wants to experiment with methane gas production. Mr. Singh explained to us that since the standard of living in rural India is far simpler than what most Americans are accustomed to, a plant of this size provides an Indian farm family with all the energy they need for cooking and lighting.

In this country, the gas produced from a continuous-feed plant the size of ours (assuming that proper care is given to the carbon-nitrogen ratio of raw materials) should be more than sufficient to meet the cooking needs of an average household, with some fuel left over for gas lights, a gas refrigerator, etc. (if you can find them). As for the slurry, every homestead worth its salt has a good-sized garden and a few acres of crops. With all due respect for compost piles, we've never seen anything else do as much for the soil as quickly as an application of bio-gas slurry.

A plant like ours is not a major feat of engineering, and we think that most experienced do-it-yourselfers should be able to build one following our instructions and diagrams. We should warn you, though, that some experience with welding and cutting metal is necessary; we used a 180-ampere arc welder and an acetylene torch.

Physically, the bio-gas plant consists of a tank to hold the wastes, an agitator, and a gas dome, which slides up or down on a center guidepipe according to the volume of gas within. Most of the materials for constructing the unit were purchased at a local salvage yard. The cost for the entire plant, including materials, some outside labor, and the excavation of the site, came to between $600 and $800.

The Tanks

For the main tank of our digester we used an old iron boiler 5½ feet in diameter and 16 feet deep. Any sturdy container of this approximate size will do as well, so long as it will hold liquid without leaking. An old gasoline storage tank would be fine. The main thing to remember if you use a different-sized tank than ours is to build your gas dome at least 6 inches less in diameter than the diameter of the tank so that the dome will not bend. If it does, the pressure of the gas will force itself out and escape into the air. Our boiler cost

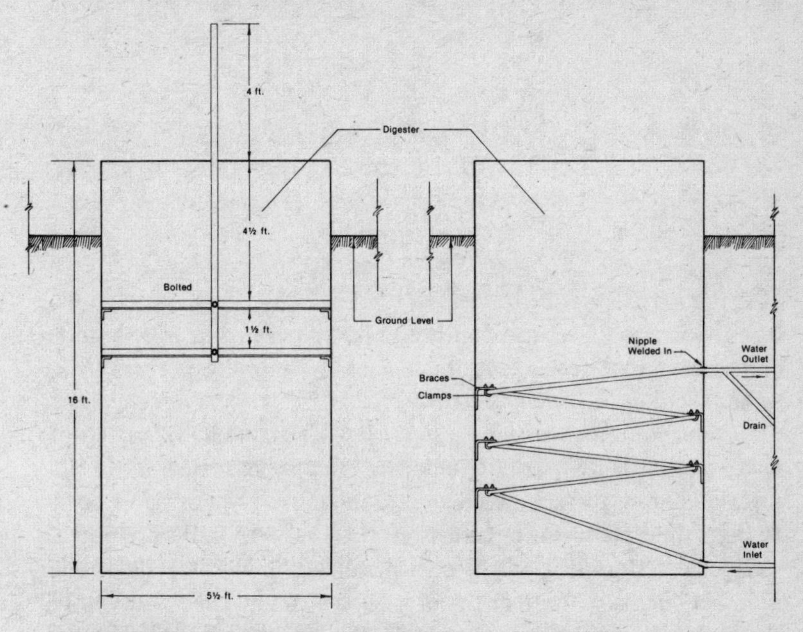

FIGURE 71. Two views of the digester tank. Left, with cross braces and center pole; and right, with heating coil.

$200 and was the largest single expense involved in the project. (Both ends were out of this boiler when we acquired it; if you find one with one end on, go right ahead, no problem.)

We welded two *cross braces* across the tank—one at a depth of 4½ feet from the top, the other at 1½ feet below that (or 6 feet from the top). We positioned these slightly 1¼ inches off center, as their purpose is to hold a *center pole* in place. The center pole

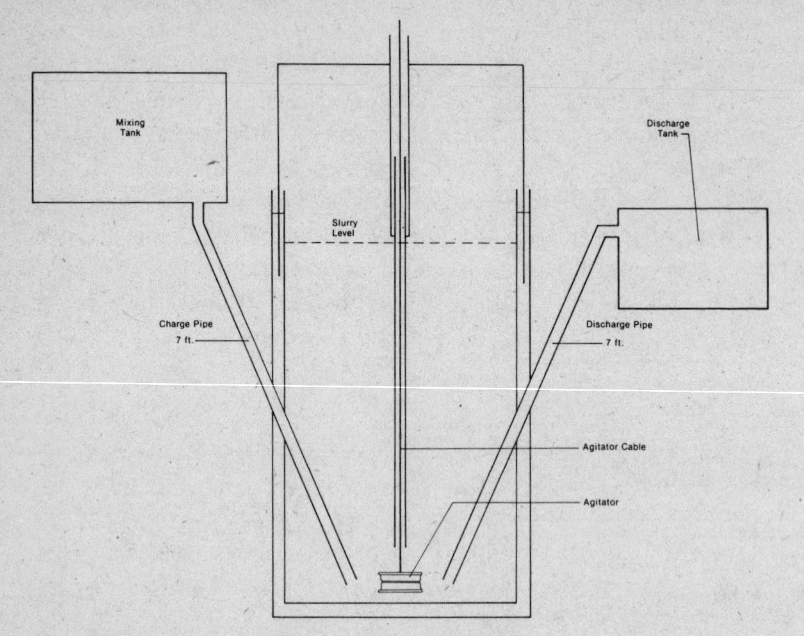

FIGURE 72. Digester with storage and mixing tanks.

is a 2½-inch pipe centered in the tank, bolted to the cross braces, and extending 4 feet above the top of the tank. (See Figure 71, right.)

Next we attached the *heating coil.* This coil allows waste hot water from bathtub, washing machine etc., to be carried out to the tank via a hose, and circulated around the tank. The hot water helps to keep the tank warm in cooler weather. (The optimum temperature for digestion is 95°F.) We then welded 6-inch braces of angle iron in a circular pattern at 6-foot intervals around the lower half of the tank to support the heating coils. We bought approximately 55 feet of 2-inch galvanized pipe for the heating coils. We had to take the pipe to a local plumber who bent it into a 4½-foot spiral form. (He wasn't too enthusiastic about this job, and it cost us quite a bit. We recommend that you investigate the types of heavy flexible plastic pipe available; something of this sort could be bent to shape at home and fastened to the braces with clamps. Although we don't have actual cost comparison figures on plastic versus galvanized pipe, we're sure the plastic would be substantially cheaper).

We set the spiral heating coil in place and clamped it to the braces, then fitted an elbow to the end of the first coil with a 6-inch fitting running out through the side of the tank. The hot water hose was attached to the bottom of the coil, as heat rises to heat the entire tank. We used the same type of fitting on the top end of the bottom coil. After the tank was in place, we dug a small ditch that acted as a drain to carry the waste hot water away from the digester. Drainage tile or stones can be used to line the ditch.

The final step in the tank construction was installing the *inlet and discharge pipes.* We cut a 4-inch circle in either side of the tank about 6 feet down from the top, and welded in a piece of metal pipe to form a collar. (See *Bio-Gas Plant* if you are placing your tank completely underground; your inlet and outlet set-up will be different.) The collar holds the two 7-foot lengths of 4-inch plastic pipe in place; these charge and discharge pipes are angled downward into the tank. After the digester was assembled on its permanent site, we arranged a mixing tank for fresh slurry on one side and a tank for catching the used slurry on the other. The mixing tank is level with the top of the digester tank, and the storage tank is positioned just under the top of the discharge pipe, so that when the plant is used on a continuous-feed basis, the system will work by gravity flow and be self-leveling. (See Figure 72.)

The Gas Dome

We found building the gas dome to be the most painstaking and time-consuming phase of the project. First we built a *frame* for the dome. We made two angle-iron hoops 5 feet in diameter by cutting small notches until the angle-iron could be bent into a circle and the ends welded together. (See Figure 73.) We then welded three angle-iron braces, each 4 feet long, to join the top and bottom hoops vertically. (See Figure 74a.) We welded small vertical braces at intervals around the edge to give the frame more support (Figure 74b). We now had a frame 5 feet in diameter and 4 feet high.

We purchased three sheets of mild steel 4-by-8-by 1⁄16 inches for the next step of the dome. We wrapped two sheets around the perimeter of the frame, outside the angle iron, and carefully welded the seams. Great care must be taken here; the steel is thin and holes can be burned into it very easily, which could cause leakage

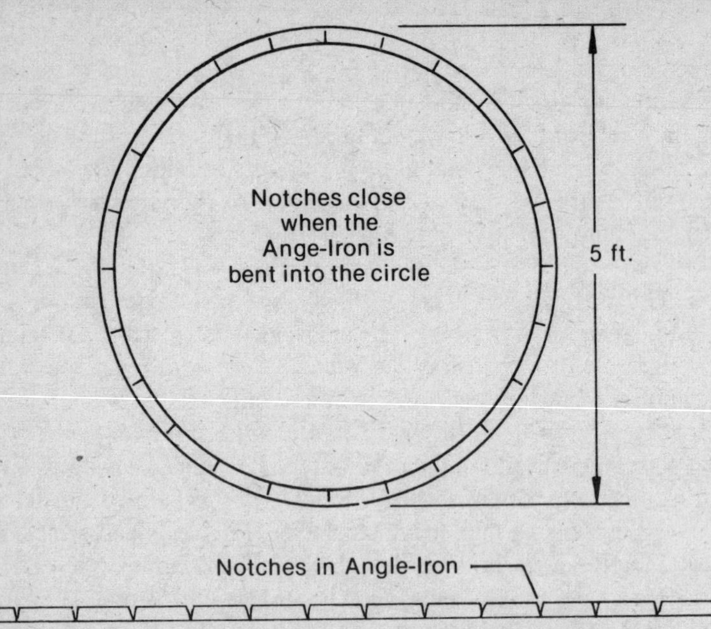

FIGURE 73. Bending the angle-iron into a circle for the hoops of the dome.

problems later on. From the remaining sheet of metal we cut a circular cap to fit over the top of the frame, then cut a 3-inch circle out of the center to receive the 3-inch-by-4-foot pipe that the center pole of the dome (as opposed to that of the tank) passes through. (See Figure 75.)

We began making the *center pole* by positioning an 8-foot-by-2½-inch pipe in the center of the dome, making sure that it went up through the 3-inch center hole in the cap and extended 4 feet above the dome. (You will note that this hole is ½-inch bigger than the center pole of the tank. This is so that when assembled, the gas dome will automatically rise and fall on the pole as the supply of gas inside increases and diminishes.) To hold the dome center pipe in place, we cut a metal plate 8 inches square by ⅛ inch thick and cut out a 3-inch circle in the center. We welded angle-iron braces from the corners of this plate to the edges of the top and bottom hoops and inserted the 3-inch-by-4-foot pipe. We then welded three small handles to the cap of the dome, but could have just as easily bolted them. Since the braces on the bottom hoop of the dome frame also serve as agitator vanes, by grasping the handle we are able to turn the dome, thereby stirring the material in the top of the tank.

The last thing to be attached to the dome was the gas valve or *faucet.* We cut a small hole about 6 inches in from one edge of the dome and inserted and sealed in place a ¾-inch piece of pipe 6 inches long. We attached an elbow to the end of this pipe and a faucet to the end of the elbow. We attached one end of a length of flexible plastic hose (left over from our milk transfer system) to the faucet and the other end to the burner of an old gas stove. (When we have visitors, we turn on the gas and light the burner to demonstrate the flame. If you were planning to use the gas for cooking in your home, you would want to substitute a more conventional hook-up.)

To finish the gas dome, we sealed the seams. Singh made a mysterious brown concoction for this purpose, but since he didn't give out the recipe, we recommend liquid steel (available at hardware and auto supply stores) daubed on the seams. When this had set, we painted the dome inside and out with four coats of enamel paint.

The Agitator

The *agitator* was really easy to make. You may improvise; we used the brake drum from an old car and attached it to a length of metal cable, about 24 feet by ¼-inch. When the tank was being assembled, we lowered the agitator to the bottom of the tank and threaded the cable up through the center poles of the tank and dome. (See Figure 72.) To move the agitator, we climb up on the dome and push and pull the agitator up and down. If we had used a cable that was long enough to reach the ground we could have attached a pulley to the center pipe so that we could pull the cable and move the agitator up and down from the ground.

Location and Assembly

The site choice on our farm posed some problems. We figured the natural place for a cow manure digester was near the source of

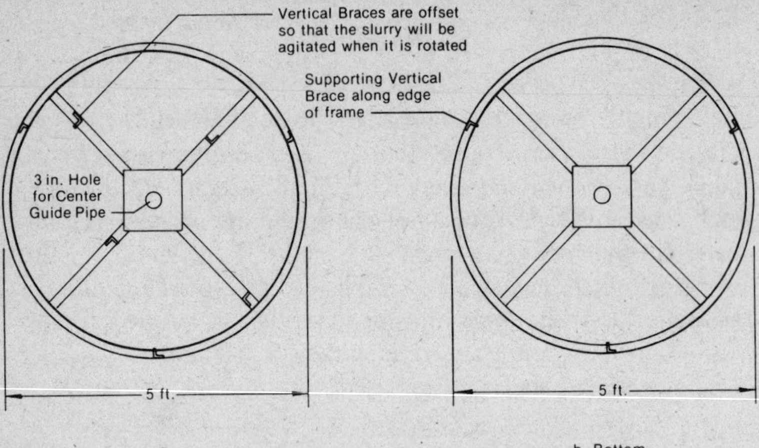

FIGURE 74. Hoops for the frame of the gas dome.

the raw material; however, Singh predicted, accurately enough, that many visitors would be arriving to see this unusual installation and that they shouldn't be expected to muck their way through the barnyard! The only possible spot near the house was on the north side. If you can arrange it, a southern exposure would provide more heat for cool weather operation. In a small set-up like ours, it is not practical to get involved in compressing gas, so the generator has to be near the point of use or you will end up with miles of copper tubing! The manure is the more moveable commodity at this point!

We hired a backhoe to excavate the site and make a pit for the digester to rest in. At 10 feet, water began to ooze into the hole, so we stopped digging. We poured a 3-inch pad of cement into the bottom of our hole, smoothed it, and let it harden. Over this we placed a sheet of black plastic (the kind used to cover trench silos), then a floor of boards, and over that, a second sheet of plastic. We carefully lowered the tank into position and sealed the whole business in place with a second layer of cement. We threaded the agitator cable through the center pipe and settled the dome over the center guide. Then we were ready to start making methane.

Operating the Plant

Several days before our digester was scheduled to start doing its thing, Singh brewed up a *starter* consisting of equal parts cow manure and warm water. He put this in a 50-gallon drum, covered it, and left it to ferment in the sun. He partially filled the mixing tank with cow manure, added an equal volume of warm water, stirred the slurry, and poured it down the inlet pipe into the digester. Then the starter was added. This "stew" is agitated 2 or 3 minutes twice a day to prevent scum from forming or heavy matter from settling to the bottom of the tank. Either condition would slow down the action of the methane-producing bacteria. Depending upon outside temperatures, gas production should begin in from three to 15 days. We knew it had begun because the gas dome began rising almost imperceptibly at first, and then more noticeably as the digester really got percolating. Because ours is a continuous-feed digester basis, no more fresh matter was added until the plant had been actively producing gas for several days.

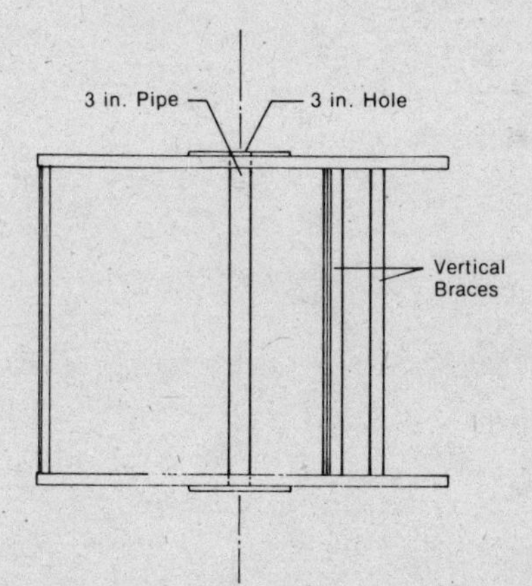

FIGURE 75. Sideview of the frame for the gas dome.

Continuous vs. Batch-Feed

The continuous-feed method of operating the plant has several distinct advantages over batch-feed use. In batch-feed, a full "charge" of all the slurry is loaded in the digester at one time, along with the starter, of course. The digester is then left alone; no more material is added. It can take three to four months, depending on weather conditions and supplied heat, for the material to be completely digested. When the methane production tapers off, the digested effluent is removed, with the exception of a small amount to be used as the starter. Then the next batch goes in.

The continuous-feed method involves more work with a small plant the size of ours, but the addition of fresh material on a regular basis means that gas production will be more even than in the batch-feed method. Small amounts of the prepared waste—half dung, half warm water—are added to the digester every day. This

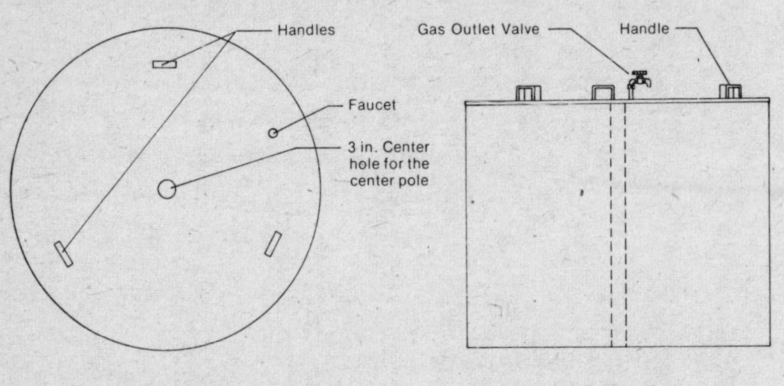

FIGURE 76. The finished dome.

insures that there is always fresh slurry in the tank producing gas full steam. Since it takes about 50 days for good digestion of animal wastes, one would need to figure roughly the amount to be added at specific intervals, so that the entire contents of the digester will have been emptied at the end of a predetermined length of time. (For example, if a digester holds 500 gallons of slurry, then to change the contents once every 50 days, one would have to add 10 gallons of new slurry each day so that 10 gallons of old slurry would be pushed out.)

Although we had first intended to continuously feed our plant, the awkwardness of carrying fresh material to it daily prompted us to revert to a batch-fed operation. We feel that gas production has been most satisfactory for our purpose of demonstration. When we eventually build our big multi-digester unit, the whole thing will be operated on an automatic continuous basis.

Temperature

Temperature has a good deal to do with the efficient digestion of the wastes. It directly affects the speed with which material decomposes and the amount of gas generated from a given batch. During the hot summer months gas production increases. For this reason it is safe to shorten the digestion cycle of a batch-fed digester by several days, or in the case of a continuous plant, to add more raw matter daily. Winter temperatures, such as are common in Vermont, slow down the action of the digester significantly, and some type of shelter and heating provision should be made. We partially buried our plant for better insulation, but even a rough wall surrounding the plant, about a foot away from it, stuffed with dry leaves, straw, or sawdust, would provide fairly good insulation. Even though digestion is best at 85–100°F, 75°F is adequate for efficient digestion.

Using and Storing the Gas

With a small plant like ours there is no problem storing the gas. The weight of the dome has been designed to force the gas out through the hose to the burner of our gas stove at the proper pressure. (For detailed discussion of domes and counter weighting, see *Bio-Gas Plant.*) You would be using the gas almost as fast as it is produced with a plant this small, so the dome is adequate for storage. We have not gotten into compressing gas at this point, but we know that it requires 1,100 pounds on a three stage water-cooled compressor to do the trick. We are undecided whether it would be most practical to compress the gas to the point of liquidation, or just to the point where it would require less space. We will be doing some experimenting with this when the large plant we will be building is completed.

Methane gas has the same uses as natural gas (i.e. cooking, lighting, heating, etc.) but minor adjustments should be made in the

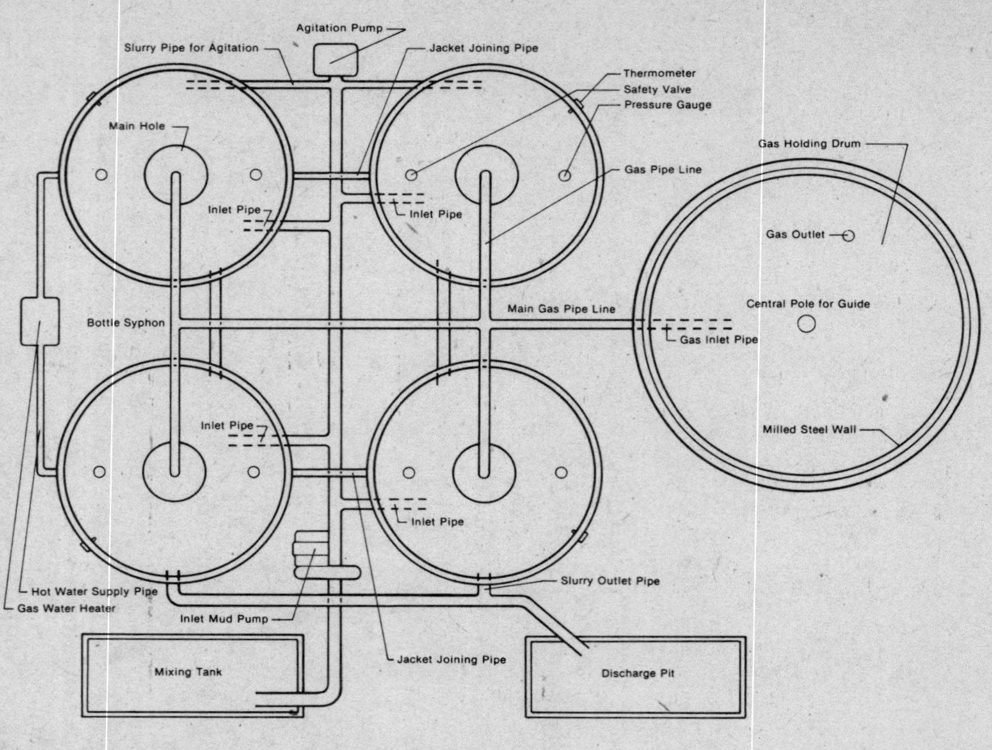

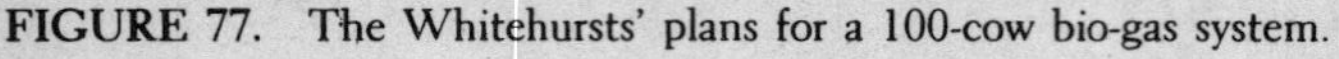
FIGURE 77. The Whitehursts' plans for a 100-cow bio-gas system.

orifice of any appliance commonly used for natural gas to insure best utilization of methane. Diesel engines run well on a fuel which is three parts methane and one part diesel. Farmers in Africa, India, and Germany have successfully used methane in stationary engines designed for gasoline or diesel fuel. Minor adjustments are necessary in the carburetor. We have run a John Deere gasoline tractor on methane, as a stationary engine. The tractor was parked by the digester, the carburetor removed, and the flexible hose from the dome thrust up into the intake manifold. The air was regulated manually. Some companies, notably Pratt and Whitney in Connecticut, are now working on the development of a fuel cell which will convert natural gas directly to electrical energy. Methane could be used in such a process also.

Using the Slurry

The digested slurry has several features which make it more desirable than raw manure as a fertilizer. The biggest advantage is the increased amount of nitrogen which is stabilized and made available to the soil by the anaerobic digestion process. Bio-gas slurry is a tremendous source of humus and therefore a real soil builder. One application of the spent slurry to our garden last spring really worked wonders. Our hard clay soil has always been a problem, muddy when wet and resembling cement when dry. After liberally soaking the garden plot with slurry, we tucked in our vegetable seeds and waited for Singh's promised miracle to happen. We weren't disappointed! The hard unruly soil became loose and friable, easily cultivated, and brought forth an abundance of prize vegetables. An added attraction of this fertilizer is that the digested slurry, unlike that familiar barnyard stuff, has no disagreeable odor and contains no pathogenic bacteria or weed seeds, these having been killed by the digesting process within the plant.

Future Plans

The bio-gas concept has potential for widespread use and can be adapted for various municipal and agricultural situations. We have plans on the drawing board for a bio-gas system which would handle the waste from our entire herd of 60 Holstein milkers and some 40 head of young stock. (See Figure 77.) This would be a completely automated system, with four large digesters and a separate gas holder. A shed would enclose the entire operation. Waste from the barn would be carried on a conveyor to a delumper-grinder unit, then into a mixing tank. There would be storage tanks for holding digested slurry through the winter months when spreading isn't practical.

We estimate that the cost of this large installation would be in the neighborhood of $20,000, and our studies lead us to believe that the value of the fertilizer and fuel equivalent of the gas would pay back this investment in three to five years. When bio-gas plants can be mass-produced, the cost of the individual installation would naturally drop.

5 SOLAR POWER

Space Heating With Solar Energy

Robert F. Girvan

In the past few years, concern over the adverse effects of burning fossil fuels has revived the age-old desire to use the sun's energy directly. Very recent increases in the cost of fossil fuels for home heating have especially served to heighten interest in using the sun's energy to heat houses.

Fortunately there have been enough solar-heated homes built in recent years to demonstrate the technical feasibility of solar heating. In this chapter I'll discuss selected features of some of these structures.

Building a complete solar-heated house is a major project, like building a house, only bigger because the solar-heating system is not a standard commercial unit. I don't expect that anyone would embark on such a project with no more background than this article, although a person might after studying some of the references given. My discussion of solar-heated homes is intended rather to provide ideas that might be adapted, on a smaller scale perhaps, to one's own home. Some ideas for doing this will be discussed toward the end of the chapter.

Heating Through a South-Facing Window

The simplest way to heat your house with solar energy is to draw back the curtains and let sunlight come in through a south-facing window. The sunlight is absorbed by the furnishings in the room and turned into heat. Unfortunately south-facing windows on most houses do not help to heat houses because they are not covered with insulating material at night—more heat may be lost through the window at night than is gained during the day. This situation may be remedied by using insulating panels at night; even heavy curtains help. See Appendix II for a good discussion of insulating windows.

The Baers' Solar House in Albuquerque, New Mexico.

If you want to obtain a substantial part of your houses' heat through its south-facing windows, then they must cover a large part of the south wall. Even then the solar heat will not be available on cloudy days unless some provision has been made to store heat from sunny days.

The Baer House (Albuquerque, New Mexico)

Steve and Holly Baer's house has 2,000 square feet of living area and all the features required for heating through a glass south wall. Almost the whole south wall of the home is covered with double-layered glass. The house is polyhedral in shape, but its unusual architecture has little, if any, effect on its solar-heating capacity. A glass south wall in any house would be just as effective.

The panels lying on the ground are insulating covers which are closed tightly over the glass walls when the sun is not shining. The inside surface of the covers is reflective (anodized aluminum). When the cover is closed, the reflective surface helps insulate the house by reflecting back heat radiation from the living areas. When the cover is open and lying flat on the ground, it reflects a little extra sunlight —about 10 to 15 percent—into the house. More information on the actual construction details of the house and on the solar energy work carried on by the Baers and their associates can be found in references 1, 2, and 3. (See Figure 78.)

Because there is so much south-facing window area in the Baer home, it might get uncomfortably hot inside on sunny days, if the sunlight were allowed to enter and be directly absorbed by the furnishings. Instead, 55-gallon drums, each with 52 or 53 gallons of water in them, are stacked behind the windows. The ends of the drums which face the windows are painted black to increase solar heat absorption.

Sun, shining on the blackened drums, heats the water in them, and when this water reaches a temperature higher than that of the rest of the house, the drums lose heat to the house by radiation and convection. In this manner the house is heated more slowly than it would have been if the sunlight had been absorbed directly by the

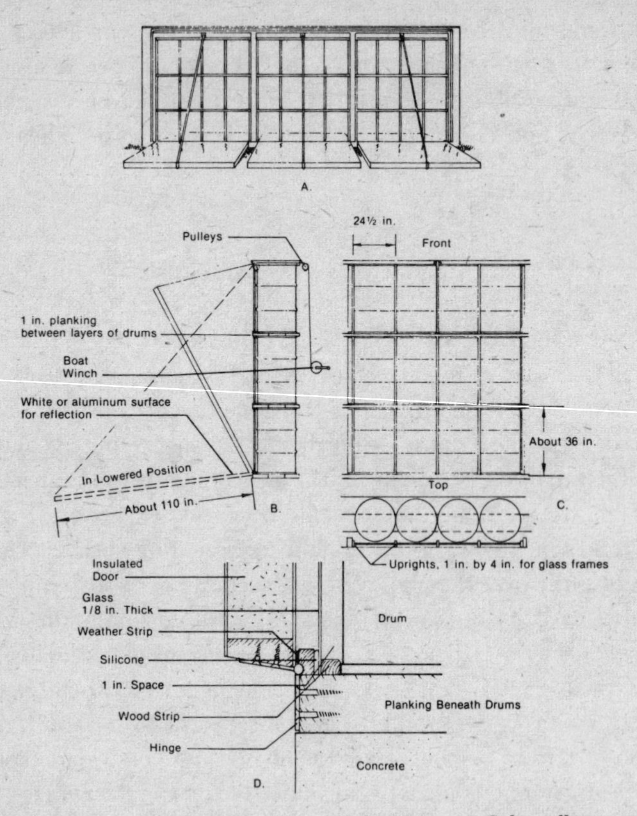

FIGURE 78. Details of Steve Baer's Solar House. A. Solar collector on the south wall. When opened, the shiny inside surface of the panels reflect the sun's rays onto the stacks of drums. B. When the sun comes up in the morning, the doors, or panels, are lowered. On very cloudy days and at night, the doors are pulled shut to prevent heat from escaping. C. Section of the south wall. D. Detailed construction of the hinged insulation panels that cover the solar collector.

interior surfaces of the house. If the house becomes too warm, the heating rate can be further reduced by closing heavy curtains between the drums and the rest of the house. The drums then continue to absorb heat from the sun, but most of the heat is stored for later use.

Even when the curtains are not closed, the drums give up their heat slowly; they stay warm after the sun sets and continue to heat the house during the nighttime hours. If the temperature of the house drops out of the comfort range, the Baers start up their wood stoves.

In the summer months, the big insulation panels over the south windows are closed and the drums have a slight cooling effect. They absorb much of the heat that enters the house, thus moderating the temperature rise. There are water containers stored at other places in the Baer home also. These, the concrete floor, and the thick adobe walls, along with the drums of water behind the windows, combine to give the Baer house a heat storage capacity which far exceeds that of the average house.

The Lefever House (Stoverstown, Pennsylvania)

The Lefevers' Solar House in Stoverstown, Pennsylvania.

Notice in the photograph of the Lefever solar house that the house is at the near end of the picture and a workshop is joined at the far end. The shape of the Lefever house is similar to that of the famous solar-heated house built by Telkes, Raymond, and Peabody in Dover, Massachusetts in 1949. The north-south dimension is much less than the east-west dimension, and the south wall is higher than the north one. This design tends to minimize that part of the surface area of the house that is not useful in collecting sunlight because it is apparently looked upon as just so much extra surface through which heat can be lost. Thus, there is a large south wall area and a small north wall area. The house is buffered against chilling winter north and north-west winds by an uninsulated storage shed built on to the north side of the house.

The large glass panels just under the roof cover the solar heat collectors. A side view of a flat plate collector, like those used by the Lefevers, is shown in Figure 79. Sunlight shines through two panes of glass and is absorbed by a steel plate with a flat black finish. Each sheet of glass on the collectors is 4-by-8-feet by-7/32 inch. The Lefevers were concerned that sunlight reflected from the large area of glass on this house might annoy their neighbors, so they used glass with a matte finish which breaks up the reflected beam.

There are no insulating covers to put over the glass panels at night. Instead the collector unit is permanently insulated from the rest of the house so that house heat can't be lost through the large sheets of glass.

When the sun is shining, the temperature of the absorber rises, and the air in contact with the back surface of the absorber is heated. When the temperature of the absorber rises above a preset thermostat temperature, ducts are opened so that fans can withdraw the heated air to the storage closets and allow it to be replaced with cooler air which is then heated. Baffles force the air current to flow over the whole back surface of the absorber.

The Lefevers' heat distribution system, including closet space for barrels of Glauber salt, is also modeled after that of the Telkes house. However, no Glauber salt was ever put in the Lefevers' closets, a circumstance that was perhaps fortunate in view of the fact

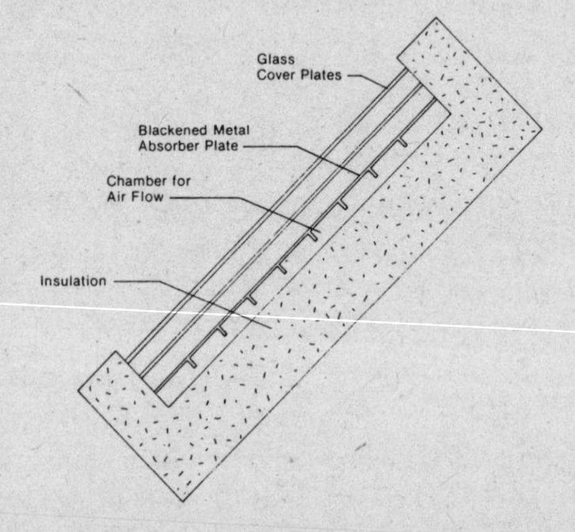

FIGURE 79. A cross-section of the type of solar heat collector used on the Lefever house.

that research is still being done to perfect this method of heat storage.

The heat distribution system is a forced air system. Hot air from the collectors is blown into the storage closets, as is shown schematically in Figure 80, a diagram of part of the system in the Telkes house. The storage closets divide first floor rooms in the house so that heated air can be blown from the closets directly to the first floor rooms as needed. Space on the second floor is used largely for storage.

There are 15 4-by-8-foot collector panels on the Lefever house, totalling a collector area of 480 square feet for a house that appears to have a living space floor area of about 1500 square feet. It has been estimated that these collectors supply something over 40 percent of the winter heat supply to the house, a percentage no doubt made possible by the fact that the Lefevers do not require their system to keep the house at a constant temperature.

Since the house has no heat storage capability beyond that provided by the heat capacity of the structural materials of the house, the solar heat must be used as it is collected. On sunny winter days more solar energy is collected than is needed to heat the house to usual living space temperatures. On these days the circulating fans could be shut off, but then the absorber plates would heat up to the point where all the heat they absorbed would be lost back to the environment. To avoid wasting solar energy in this way, the fans are left running, and the house is allowed to continue heating. The Lefever family merely changes to lighter clothing. The well-heated house presumably then stays warm into the night. (This willingness to adjust to the daily variations in the supply of solar energy by putting on or taking off clothing would seem to simplify considerably the task of designing solar houses.) On days when there is not enough sunshine, the house temperature is allowed to drop below normal. However, there is a fireplace and other auxiliary heat which can be used to keep the house from becoming really cold.

Steps have also been taken to keep the Lefever house cool in summer. There is a prominent roof overhang on the south side of the house which partly shades the collectors. At the bottom of the collectors there is a smaller overhang which projects far enough to shade the house windows in summer.

There are many other types of air-heating collectors besides

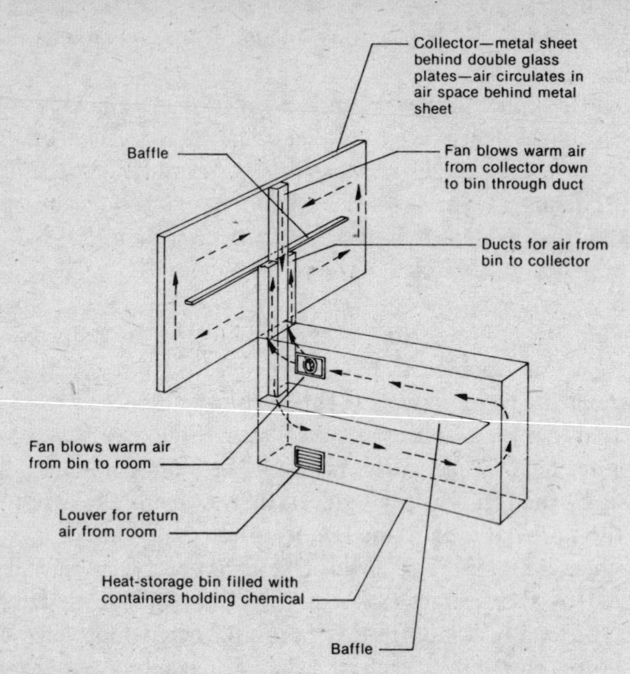

FIGURE 80. The heat distribution network for a two-collector subunit of the Telkes solar house, built in 1949 in Dover, Massachusetts.

those used in the Lefever house. In some a louver construction is used for the absorber, the heated air flowing between the slats. In others, blackened gauze or expanded metal, like that used in oil filters and air cleaners, is used to absorb the solar energy and transfer it to air flowing by. Discussions of these and other types of collectors can be found in *Direct Use of the Sun's Energy*, by Farrington Daniels. (See Bibliography.) The Daniels book contains good information on all aspects of solar energy utilization and is a valuable book for any person interested in this field.

The Last M.I.T. House (Lexington, Massachusetts)

The solar heat collectors on this house, which was built by the M.I.T. solar energy group in 1958 (and was later dismantled), were mounted on the south roof, tilted 60° from the horizontal. Because of the tilt the collectors were aimed almost directly at the noonday sun in midwinter. Their surface area is 640 square feet. The house had two floors of living space totalling 1450 square feet and was well insulated, as all solar houses must be. During the house's operating years the heat loss was only 30,000 Btu/hour when the outside temperature was 0°F.

The solar energy collectors in this house heated water rather than air. Copper tubes were attached to a blackened aluminum absorber plate, and the water to be heated was pumped through these tubes. A sketch of this type of flat plate collector is shown in Figure 81. In an earlier M.I.T. house, a blackened copper sheet 0.02 inches thick was used as the absorber. Copper tubing which had an outside diameter of 0.375 inches and a .016-inch wall thickness was soldered to the absorber with a 6-inch spacing between lengths of tubing. The collectors in both the earlier and last M.I.T. houses were small and lightweight and had a small heat capacity so that they could respond rapidly to changes in solar energy, such as on partly cloudy days.

The overall efficiency of the collectors was 43 percent over a two-year period. Energy loss is caused by the fact that some of the incident energy is reflected and some of the heat that is collected is lost to the surrounding area.

The last M.I.T. solar house—built in Lexington, Massachusetts in 1958.

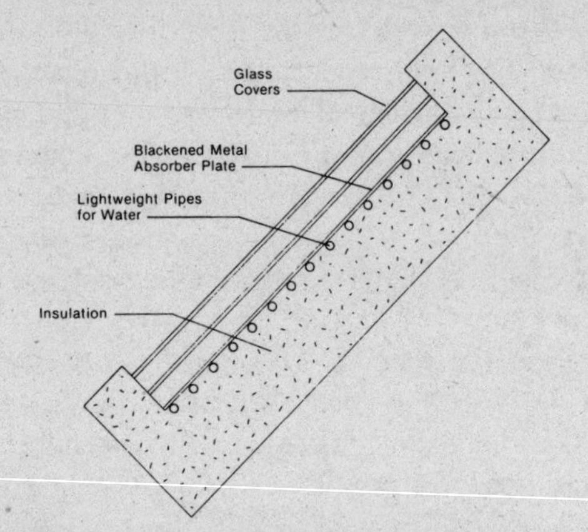

FIGURE 81. A cross-section of the type of flat plate collector used on the last M.I.T. solar house.

The storage system for the last M.I.T. house consisted of a 1500-gallon tank of water. With this storage volume and collector area, the solar heating system supplied about 48 percent of the house requirements during two winters, the rest being supplied by an oil furnace. This combination of oil heat and solar heat was deliberate, as it was thought to be the most economical way of heating the house at the time. I have taken most of the data on this house from reference 5, which provides more extensive discussion of this house, including a schematic diagram of its heating system.

Thomason House No. 3

In the Thomason residence in Washington D.C., the rooms which are heated by the sun have a floor space of 1500 square feet. The solar heat collector has an area of 960 square feet.

Thomason's first solar house was built in 1959. Its absorber was black corrugated aluminum and the corrugations ran from top to the bottom of the roof. Water was pumped to a ½-inch copper tube which ran along the peak of the roof. A hole was drilled in the pipe over each valley in the corrugated aluminum so that the water could dribble down the valley, absorbing heat from the aluminum in the process. The heated water was collected at the bottom of the roof in an insulated gutter and pumped to a storage tank.

Each succeeding Thomason house has improved on the original design. In house No. 3 the glass covers over the absorber are attached with aluminum frames rather than with the redwood strips in house No. 1. Unlike the first two houses, which have their collectors at the bottom of the roof, house No. 3 has its collectors up on the roof where they are less susceptible to shading from nearby trees and buildings.

The living space of all three Thomason houses is heated by a forced-air system. To transfer the solar heat from the water in the collectors to the air that is to be circulated through the living space, Thomason uses a bin of fist-sized rocks around the water storage tank. The rocks are directly in contact with the outside of the storage tank so that heat from the water inside the tank can heat the rocks.

When the temperature in the living space drops to that set on a thermostat, an electric blower is turned on to draw cool air from

The Thomason solar-heated residence in Washington, DC.

the living quarters through the rock bin. Because the rocks are of uniform size there are no small ones to plug the spaces between bigger ones. As the air meanders through these spaces, it is heated; then it is returned to the living space.

The rock bin serves two purposes. First, it provides a large surface area to transfer heat from the tank of heated water to the house's forced-air heating system. Second, it and the water tank store heat for cloudy days. In house No. 3 the masonry water tank holds 3,000 gallons of water and the rock bin holds 25 tons of stones.

With the given collector area and storage capacity, Thomason estimated that the solar-heating system for house No. 3 could provide 75 percent of its winter heat. In fact, the roof of the indoor swimming pool (the one-story porch at the left in the picture) is covered with an aluminum reflector. Sunlight reflected from this nearly horizontal roof increases the performance of the system to such a degree that one winter the house was heated 90 percent by the sun and 10 percent from an auxiliary furnace.

More information on the Thomason houses can be obtained from his many publications.[6–9]

Other Designs

Zomeworks Corporation has constructed a solar-heated house in which air circulation in the collector takes place by natural convection.[10] However, the house must be placed above the collector and the collector occupies enough of a vertical dimension to essentially require that it be on the side of a steep hill or mountain.

Harold Hay has designed another solar-heated house that is especially suited for hot, dry regions.[11] His one-story house makes use of the heat capacity of a volume of water stored on the roof to heat the house in winter and cool it in summer. The water is stored in plastic bags which have been secured between beams above the ceiling. Insulation panels that move on aluminum tracks have been placed over these water-filled bags to insulate the water from the cold night air in winter and from the hot daytime sun in summer. The interior temperature of his house stays between 62 and 82°F, despite the fact that outside the temperature ranges from below freezing to 115°F.

I've only discussed a few of the dozen or more solar-heated houses in the world to demonstrate the different ways that solar energy can be collected, stored, and transferred to living spaces. Additional information on these and other houses and on the principle of solar heating can be found in reference 5. Farrington Daniels' book also discusses solar house heating and suggests further reading on the subject.

Comparison of Selected Design Features

Because of differences in lifestyle, taste, skills, site characteristics, and available materials, there will be individuals who will prefer one of the designs that I have discussed over all the others, despite particular performance advantages of the rest. My particular preference happens to be a system like the Baers'.

I like to lump together the systems in which air or water must be circulated past the absorber under the classification "systems which use heat transfer fluid." The fluid which carries heat away from the absorber is air in the Lefever house and water in the M.I.T. and Thomason houses. I call the system in the Baer's house a "direct radiative system" even though some of the heat is carried away from the barrels by air currents and not by radiation.

The systems which use heat transfer fluids have the advantage of a positive means of transferring heat to remote corners of the house—just pump hot air or water to that point. Perhaps this could be done with the direct radiative system, too, if one wished to spoil its elegant simplicity.

The designs which use heat transfer fluids may have another advantage in that they can be shut down by thermostats in cloudy weather or at night. The Baers' system is shut down by closing a huge cover. Perhaps many of us wouldn't mind opening and closing an insulating cover each day, but for those who might, an automated alternative is apparently available. In the January 1974 issue of the newsletter of the U.S. section of the International Solar Energy Society, Zomeworks Corporation announced that a complete greenhouse had been equipped with the automatic insulating device that Steve Baer mentioned in reference 2. The insulation consists of styrofoam beads which are blown into the space between inner and outer glass panes. They are automatically removed when the sun shines. Harold Hay, who has worked with moveable insulation panels for a number of years, has also equipped his solar house at Atascadero, California with an automated system.[12]

One disadvantage of the systems which use a heat transfer fluid (besides the fact that some of them are expensive) and of the direct radiative system of the Baers' house is that most of them use a lot of glass for a design that permits the solar energy to be used only as heat. The sun's rays are more valuable to me as light than as heat. I have seen one design which uses a louvered absorber with a transluscent backing to allow light to pass through, but I believe it is still in the experimental stage.[13]

Heat Needs and Environmental Costs of Solar Houses—Some Thoughts Before Building Your Own

There are two aspects of solar house heating that are often not discussed in publications of this nature even though they are important to someone who is thinking about building a solar-heated house. One is the amount of fuel energy that is saved by heating a house with solar rather than conventional heat, and the other is the amount of fuel that it takes to manufacture the materials for the solar-heating system.

For an example of the fuel savings we will use the last M.I.T. house because complete figures are available for that unit. The average winter heat load for this house (averaged over two winters) was 88.5 million Btu's. This can be expressed in terms of fuel consumption with the aid of the following table of approximate heat values of common fuels:

furnace oil	140,000 Btu/gallon
liquefied propane	100,000 Btu/gallon
natural gas	100,000 Btu/hundred cubic feet
fire wood	20,000,000 Btu/cord

When the fuel is burned in a heating stove, only a fraction of the stored energy, perhaps 50 percent on the average, actually ends up heating the house. The rest goes up the chimney as hot air or as incompletely burned compounds. Expressed in these terms the M.I.T. solar house would have required the following amounts of fuel to heat it all winter if no solar energy had been used at all (and if its furnace were 50 percent efficient): 1,260 gallons of furnace oil, 1,780 gallons of LP gas, 178,000 cubic feet of natural gas, or 8.8 cords of wood.

In terms of the average price of 30¢ per gallon that was charged for LP gas around here in winter 1973–74, it would have cost about $535 to heat the house without any help from the sun. In terms of the 10¢ per 100 cubic feet that my mother-in-law paid for her natural gas that same winter it would cost $178.

The solar unit contributed an average of 43.5 million Btu to the house during each of two winters. In terms of LP gas at 30¢ a gallon this is worth about $260. In terms of natural gas at 10¢ per 100,000 Btu it's worth about $87.

According to Makhijani and Lichtenberg[14] about 12,300 Btu are required to make a pound of glass. The window glass I have weighs about 1½ pounds per square foot, so about 18,000 Btu were required to make a square foot of it. Two sheets of glass are required for each square foot of collector and about 17,000 Btu might be required to make a square foot of absorber. In all, perhaps 53,000 Btu might be required to manufacture a square foot of solar heat collector. The collectors on the last M.I.T. house delivered 136,000 Btu/square foot/winter, so they repaid the energy that was spent making them in less than one year. However, poorly designed units, or units in areas where the climate is very unfavorable for solar heating, or small-scale devices which are only used intermittently might require a considerably longer time to repay the initial investment of energy.

The Boston area, which was the location of the last M.I.T. house, is not a particularly favorable location for solar heating (although fuel prices do tend to be higher there). It is likely that well-designed solar-heating units would repay their energy investment in an even shorter time in much of the rest of the country. Table 29 in Appendix I gives you an idea of how favorable the climate in various parts of the United States is for solar-house heating.

The South-Facing Window as a Solar Energy Device—My Own Choices

My own priorities are governed by my situation. My family and I live on the edge of a little country town. We're the only people for miles around who burn wood, so whenever an ice storm breaks branches or a tree blows over, we get the wood. For kindling we use scraps from the lumber yard which would otherwise be burned as rubbish. The houses out where we live are far enough apart so that smoke from our chimney doesn't bother anyone.

Because of our ample supply of firewood there is not an immediate need for us to switch to direct solar heating. We have time to do it in steps, experimenting along the way. It's fortunate that this is the case since we don't have the time or the money to carry out a once-and-for-all changeover to direct solar heating.

We began making better use of the sun by treating our south-facing window as a solar energy device. On a clear December day here in Iowa the sunlight which comes through each square foot of such a window as part of the direct beam will have a heat value of about 1100 Btu. There is some heat lost from our living areas through the window during the day, but this is almost balanced by the energy income from diffuse sky radiation.

In one room we have two windows side by side to which we have paid special attention. We're trying out ideas on these windows that we'll use later on a larger scale. The windows have a total area of 18 square feet, so on a clear December day they will admit about 20,000 Btu of direct beam solar energy. This is the heat that would be obtained by burning, at 50 percent efficiency, a block of wood 7.7 inches on a side. It is not much energy, but it's about as much as would be collected by any solar energy device with the same area.

The most important function of the window is to provide a visual link to the out-of-doors. Its next fuction is to admit light to the interior. In terms of replacement cost, the sunlight that comes in the window is far more valuable as light than as heat. It would take over 500 watts of fluorescent lamps burning for 8 hours to produce as much light as the direct beam sunlight which comes through a square foot of window. Seventeen cents worth of electricity would be required to provide this light, whereas ⅔¢ would buy enough propane to supply the heat that that amount of sunlight yields.

Unfortunately, the light is not distributed evenly about the room. Near the window there is more light than is needed for adequate vision, whereas farther away the lighting may be insufficient. We plan to distribute the light better using the relflectors that I describe below.

There really isn't room for insulating panels on the inside of our windows so we plan to put them on the outside, like on the Baers' house. (We do close curtains over the window at night, though.) The panels will have reflective inner surfaces, and the first one will be hinged at the bottom so that sunlight can be reflected from its inner surface to the ceiling of the room. (See Figure 82.) The ceiling is a white, diffusing surface which will reflect the light about the room.

For a reflective surface on the panel, we'll use the aluminum sheets that can be purchased inexpensively from printing plants around here. The surface of these sheets is not as smooth as that of a real mirror so they won't reflect in as much sunlight as a really good

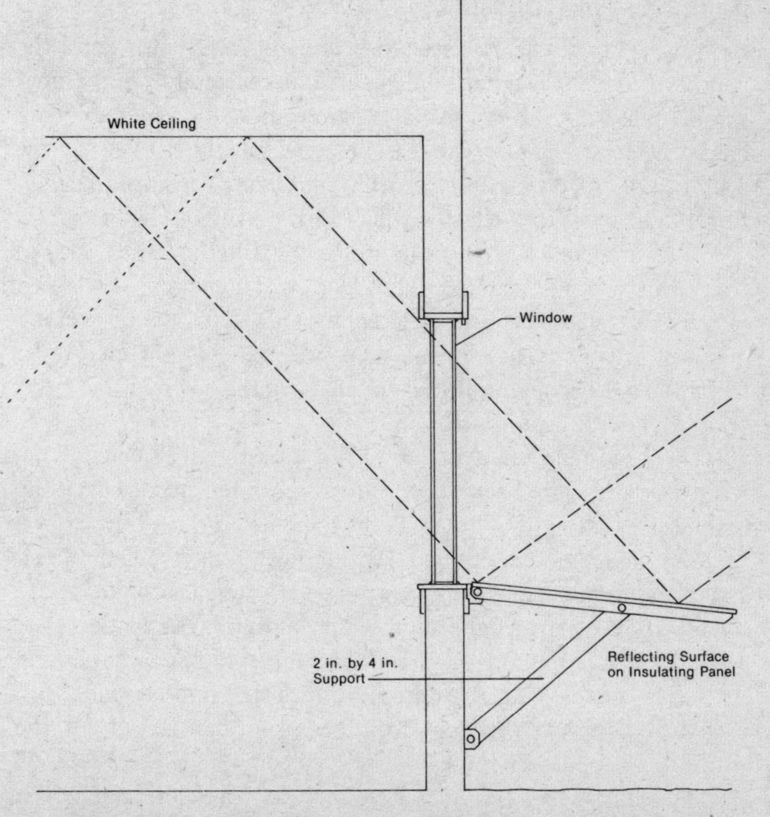

FIGURE 82. Light reflected to the ceiling of a room from the reflective inner surface of a moveable insulation panel.

surface, but this isn't a serious drawback. Even a good fraction of the direct sunbeam provides a lot of illumination. Besides, the sunlight will be incident on the aluminum at a large slanting angle. At such an angle, surface roughness doesn't reduce reflectance as much as for perpendicular incidence. The advantage of the aluminum sheets is that they're inexpensive, and if they're attached to prevent the wind from getting under them, they're weatherproof.

The light from the lower reflector will also help to dry the clothes more evenly on our window laundry dryer. (See Figure 83.) At present the shaded parts of the laundry up by the rack don't dry as fast as parts in direct sunlight, so the solar drying is not of much use; we've got to wait until the wet spots near the top dry up anyway. The reflected light will make these now-shaded parts dry faster.

The second reflector-insulator will be hinged at the top of the window. When it is open, sunlight will be reflected from its inner surface onto leafy vegetables that we grow inside the window. There has been a tendency for some of these, especially the lettuce, to grow spindly, presumably because of insufficient light. We hope the additional light from the reflector will help the plants grow better.

We've spent a lot of time fighting aphids on the lettuce the last two winters, so the economics of indoor plant growing haven't been favorable for us yet. However, judging by the harvest from behind even our present small window, we expect the indoor garden to be by far the most economically favorable use of the incoming sunlight, once we lick the aphid problem.

The two reflectors will also increase by about 25 percent the solar heat through the window. There are three reasons why the extra energy intake is so small. First, the reflectors must be set at a slanting angle to the sunlight so they won't intercept as much sunlight as they would if they were more perpendicular to the incoming beam. Second, the light from the reflectors hits the window glass at a rather large slanting angle so that a lot of it is reflected by the glass rather than being transmitted through the window. And third, except at noon, all the light from the reflectors doesn't hit the window; some hits the house wall instead.

In addition to helping us heat the house in winter, there is a fringe benefit associated with the top reflector: It can be lowered from its winter position to shade the window from the summer sun and help to keep the house cool.

The South-Facing Greenhouse and Wood Drier

We've also increased the glass area on the south side of our house by enclosing our porch with glass. For $10 I was able to buy a trailer load of storm windows and screens from a house that was being equipped with combination units. The storm windows became available on a raw winter day, and I didn't have any place to store them. My hands were numb with cold so I put the windows up as quickly as possible, using the simplest construction I could think of. In retrospect I'm glad things happened as they did. Given more time I might have designed a lot fancier porch that wouldn't have been any more functional than the one we have.

I drilled four holes through the frames of each matching set of window and screen so that they could be bolted together. Then I built a frame of scrap 2-by-4's between the posts of our south porch

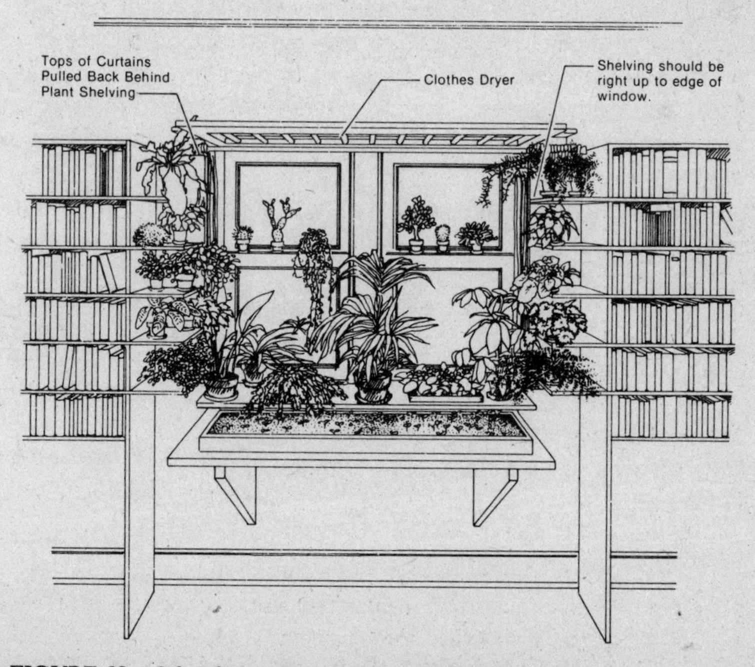

FIGURE 83. Solar clothes dryer inside south-facing window.

so that there would be a 2-by-4 behind every line where window frames would meet when they were mounted to enclose the porch. (See Figure 84.) I nailed the screens to this frame; they'll keep the flies out next summer. Meanwhile they offer some protection against children falling through the ceiling-to-floor glass. I bolted the storm windows to the outside of the screens. Now our porch has over 100 square feet of single glazed glass area.

On windless and sunny winter days the temperature on the porch sometimes reaches 85°F (or about 50°F above outdoor temperature). Next winter when we're sure we've plugged all the leaks and conquered all the drafts (by tacking felt stripping under all the screens and storm windows), we'll install a duct to transfer heated air from the ceiling of the porch to an upstairs bedroom.

Meanwhile, the porch will be used to extend our vegetable growing season. Since we've kept aphids alive over the winter in our indoor garden, we're reluctant to start spring garden plants in the house, so we'll start them on the porch. This fall we'll grow plants on the porch as long as they'll survive there. If there's time, and materials become available, we'll put hinged covers on the porch to insulate it at night and extend the growing season even further. (By the way, we will have to take great care to see that there is a tight seal around the insulating panels. Otherwise air currents flowing between the window and the panel will largely negate the insulating function of the cover.)

Last winter the sun porch was also used to melt ice from firewood. When we arrived here last fall we had time to put in a late garden, cut a wood supply and construct a brick chimney, but we weren't able to provide shelter for the wood before an early ice storm coated it. After the porch was enclosed, we kept it filled with firewood so that the ice could melt on sunny days.

Computation of Heat Collecting Potential of a Device

From my point of view it is far better to be able to take advantage of scrap construction materials as they become available and to adapt general solar-heating techniques to to one's own particular needs than to follow a fixed design. Fixed designs, like the ones I've discussed earlier, may sound very attractive, but you may find that they are expensive and difficult to adapt to your own house's structure. Alterations may be necessary to fit your own particular needs. However, before straying too far from proven designs it's a

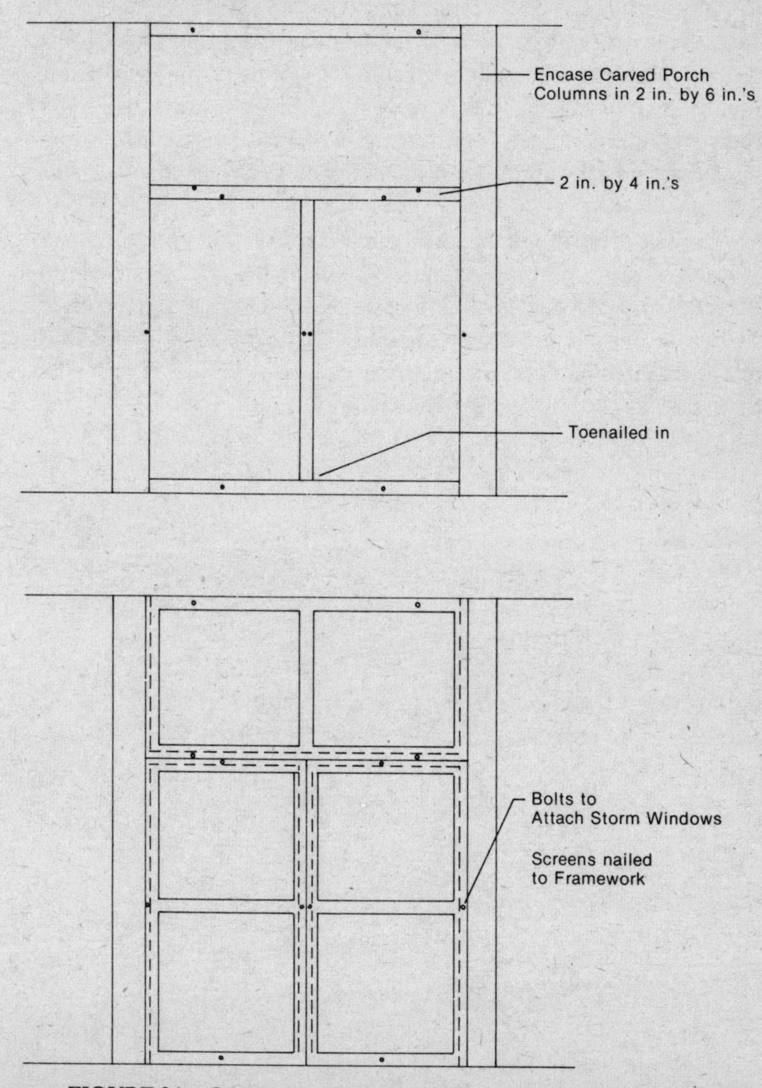

FIGURE 84. Construction for a quick and easy method of enclosing a porch with storm windows.

good idea to cultivate an ability to estimate the heat output of a new design. I prefer to do this mathematically from a few basic principles rather than from collections of tables which may be hard to find and incomprehensible when you do find them. Those who wish to follow a tabular approach will find the necessary information in Table 29 in Appendix I. (Also see reference 15.)

As an example of a case in which a deviation from tested designs might be worthwhile, I'll consider our own front porch. The roof leaks and I'm going to have to shingle it next summer to prevent the structural members from rotting. The roof faces south, so perhaps I should instead put a solar energy collector on it to keep the rain off.

My interest in putting a solar collector on the porch roof was aroused by a very simple design for a solar heater, apparently invented by Buck Rodgers of Embudo, New Mexico.[16] It's a convective air heater (Figure 85) of the same type used by Zomeworks in reference 10. The difference is that the unit in Figure 85 is smaller than Baer's unit; it fits on a house window.

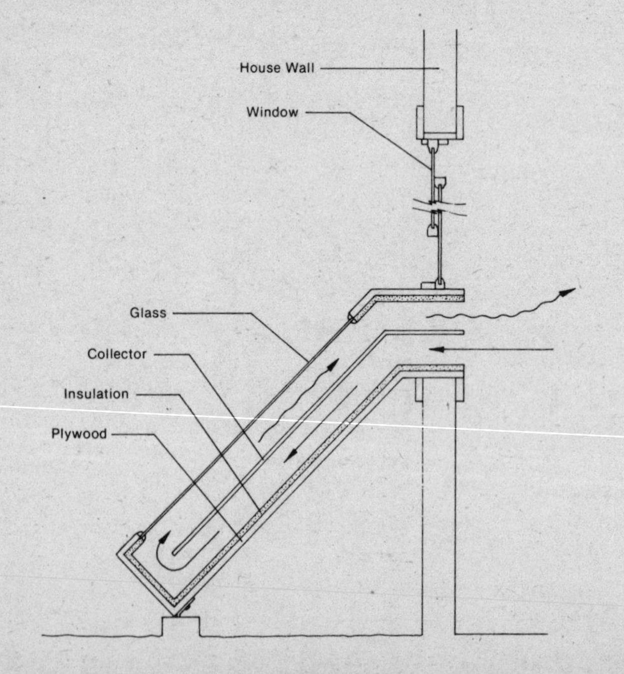

FIGURE 85. Window-sized convective air heater.

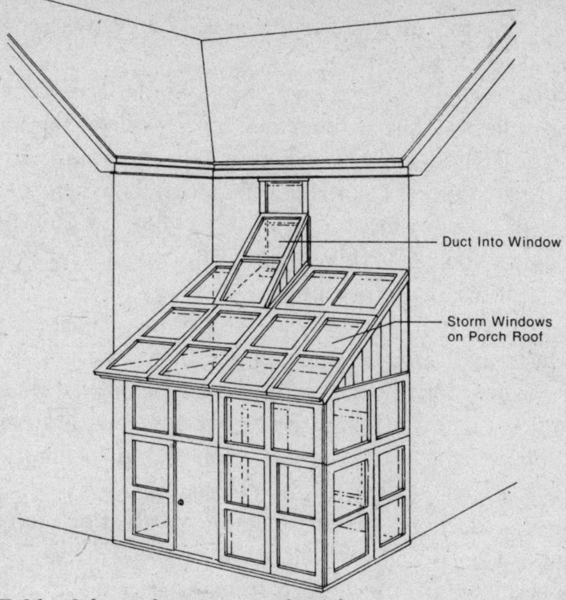

FIGURE 86. Solar air heater on porch roof.

I probably wouldn't want the window heater on a ground floor window as is shown in the figure, because I'd have to remove it every spring to make room for cold frames along the south side of the house. However, on the porch roof, where it could stay forever, such a device might prove quite useful, especially since the window just above the porch roof (Figure 86) has a storm window with panels that can be raised to make room for the window heater without tearing anything up.

After building the porch, I still have a number of storm windows left, but if I use them to make a collector, I won't have enough for a greenhouse. Which should I build? Let me take you through a series of calculations to show you how I would determine the amount of heat I could expect from my porch-roof collector.

I'm skeptical about the effectiveness of such a collector because the porch roof is too flat. I'm afraid that one installed there wouldn't collect an appreciable amount of heat. It is only tilted from the horizontal by about 23°. Since the wintertime sun is only about 25 to 35° above the horizon even at noon where I live at 42.3° N latitude, the incident sunlight will be 32 to 42° away from being perpendicular to the surface of the collector, even at noon. At other times of the day the angle of incidence will be even larger because the sun is lower in the sky.

Such large angles of incidence would hinder the performance of the heat collector. This is the reason for the rule of thumb which says that a solar heat collector which is meant for wintertime use should be tilted from the horizontal by an angle which is equal to the latitude of the site plus about 15°. When tilted at such an angle, which would be about 57° where we live, the collector is perpendicular to the noontime rays in winter.

The easiest way to find out the effect of the angle of incidence is to go ahead and carry out the computations for the porch roof. The calculation at solar noon is the easiest, since at that time everything points straight south and there's not the complicating factor of the east-west angle of the sun that comes up in the morning and afternoon. We'll only do noontime calculations.

Figure 87 illustrates the first deleterious effect of a large angle of incidence. That is that a collector which is tilted away from the sunlight doesn't intercept as much sunlight as one that is perpendicular to it. The line AB in the figure is a cross section of a collector lying on the porch roof. AB′ is a cross section of the same collector tilted up so that it is perpendicular to the rays of sunlight. Some of the rays of sunlight which strike the collector in position AB′ miss it completely when it is lowered to AB. In fact, the ratio of the amount of sunlight which strikes the collector in position AB to that which strikes it in position AB′ is equal to the length of AC divided by the length of AB′, as can be seen from the figure. By actual measurement of AC and AB′ this ratio is found to be .83 for Figure 87.

The second bad effect of having the sunlight incident at a large angle comes about because of the tendency of common surfaces to reflect light at large angles of incidence. This affects both the transmission through the glass covers (in the Iowa climate two covers should be used rather than one as in the figure) and absorption by the absorber. When sunlight hits the glass at a large angle, more of it is reflected so the amount transmitted is reduced. The absorber also reflects more of the light that hits it at large angles of incidence so there is less absorbed.

The magnitude of these two effects can be found from the graphs of Figures 88 and 89. Figure 88 gives curves for the fraction

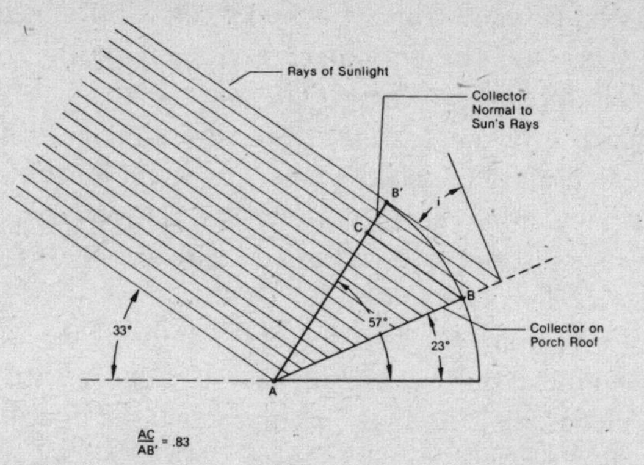

FIGURE 87. Sunlight intercepted by a solar collector in two orientations. In position AB it intercepts only a fraction (AC/AB') as much solar energy as it does when it is perpendicular to the sun's rays, as in position AB'. The angle between the rays of sunlight and a line perpendicular to the collector, labelled in the figure, is called the angle of incidence.

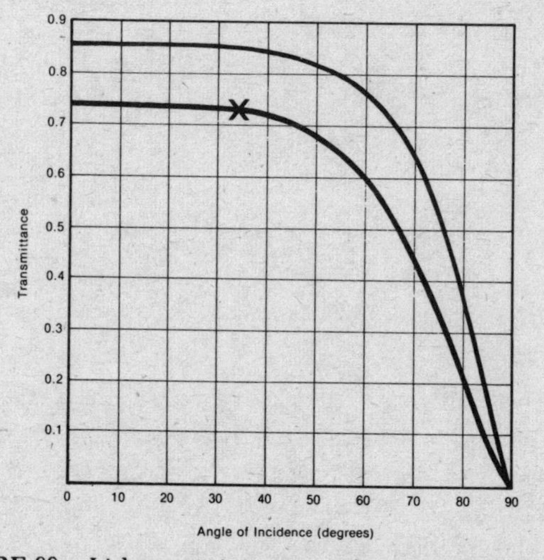

FIGURE 88. Light transmission curves for one or two sheets of glass. The quantity plotted is the fraction of light transmitted vs. the angle of incidence.

of light transmitted through glass as a function of the angle of incidence. The curves apply to sheets of glass which are ⅛ inch thick, have an index of refraction of 1.5, and absorb about 7 percent of the energy from a beam which passes through them. This is intended to approximate ordinary window glass.

The top curve gives the fraction of light transmitted through a single sheet and the bottom curve represents the light that is transmitted through two sheets. The point at an angle of incidence of 34°, which is the angle of incidence in Figure 87, is represented by x. The transmittance at this point is 0.73, not much less than that at normal incidence. However, when the sun is lower in the sky in the morning and afternoon, the angle of incidence is larger and the transmittance falls off rapidly.

Figure 89 shows the variation of the absorptance of sunlight by a blackened metal sheet versus angle of incidence. The absorptance at an incidence angle of 34° is .93, not much different from that at

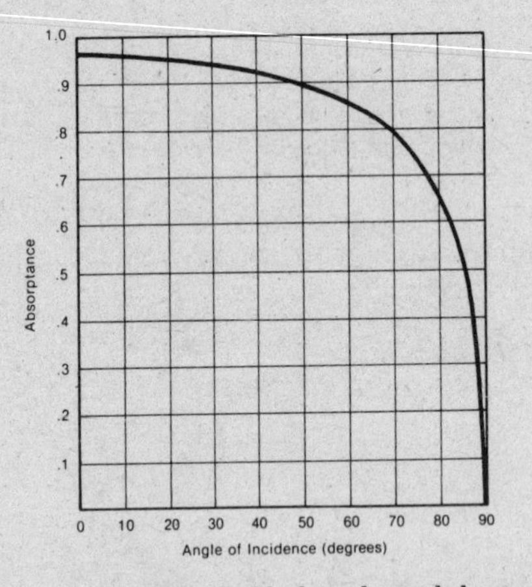

FIGURE 89. Absorptance of a blackened metal sheet vs. the angle of incidence.

normal incidence. But again, the absorptance drops off in the morning and afternoon hours because of the larger angle of incidence.

Multiplying the transmittance by the absorptance by the effective area for intercepting solar energy yields $0.73 \times 0.93 \times 0.85 = 0.576$ for the fraction of the solar energy that will be absorbed compared to what would have been incident on the outer glass surface of the collector if it had been perpendicular to the sun's rays.

The only quantity which remains to be determined is the energy in the sunbeam itself. The amount of energy depends on atmospheric conditions and on the height of the sun in the sky. Threlkeld and Jordan have compiled convenient charts which give the energy in a beam of sunlight at different times of the day and year. Their

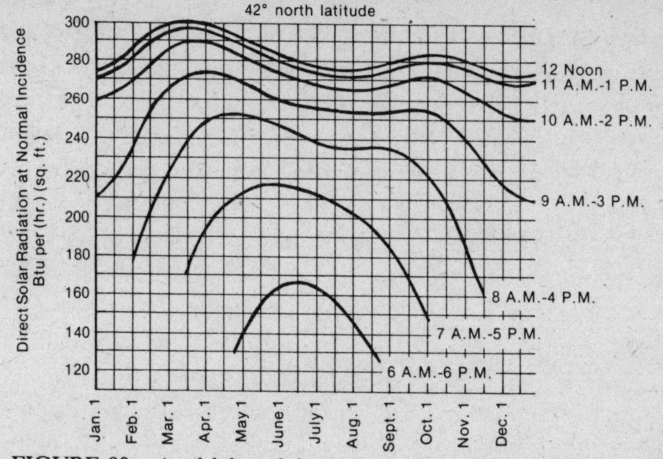

FIGURE 90. Availability of direct beam solar energy at 42°N latitude on average clear days. The chart indicates the energy per hour that falls on a square foot of surface oriented perpendicular to the sun's rays.[17]

chart for 42° N latitude is reproduced in Figure 90. Other charts at 6°-intervals of latitude can be found in reference 17, but if you're not interested in accuracy to better than 10 percent, which I'm not at present, it's all right to use this chart for the entire United States.

Figure 90 indicates the solar energy per unit time which would fall on one square foot of collector area held perpendicular to the incoming rays of sunlight on an average clear day. The chart is for non-industrial areas and is for direct beam energy only. Skylight is not included.

Threlkeld and Jordan have devised a set of clearness numbers to account for the fact that air clarity on an average clear day varies from one part of the country to another. Technically the clearness number for each locale should be multiplied by the number found from Figure 90 to find the incoming solar energy, but in most parts of the United States you'll only be inaccurate by 5 percent at the most if you don't multiply by the clearness number. In Iowa the number is 1.00 anyway.

In Figure 87 the angle between the sunlight and the horizontal, or the altitude of the sun, is 33°. You can always find the sun's altitude at solar noon from the equation $a_n = 90° + D - L$. The sun's altitude at noon is a_n, L is the latitude, and D is the sun's declination (or, in other words, the latitude of places where the sun is directly overhead at noon on that day). If we use the above equation and substitute 33° for a_n and 42° for L, we can solve for D. When we know the value of D, we can determine the date by using the *Old Farmer's Almanac* (the declination for each day of the year is given in the almanac at the top of the calendar pages). Remember that the declination is negative in the winter (the sun is south of the equator).

According to Figure 90 the direct beam sunlight at noon strikes each square foot of a collector that is perpendicular to it with 292 Btu/hr. of energy. From our calculations (transmittance × absorptance × effective area) we found that each square foot of absorber in the porch roof collector would absorb 0.576 of the solar energy falling on it. So, we see that our collector will absorb .576 × 292 Btu/hr, or 169 Btu/hr. However, not all 169 Btu/hr will be transferred to the house. Some of the heat from the collector will be lost to the outdoors through the glass. The rate at which the heat loss occurs depends on the temperature of the absorber, the temperature of the surroundings, and the windspeed. The rate increases as the difference between the two temperatures increases, or as the wind speed increases.

Let's assume that the collector is rigged so that hot air is taken from it fast enough to keep its temperature from rising above 100 °F, that the outdoor temperature is 30°F, and that the wind velocity is 10 mph. Under these conditions about 41 Btu/ft²/hr would be lost from the absorber back to the environment, leaving a net of 128 Btu/hr/ft² to be delivered as useful heat.*

---

*I haven't included the heat transfer curves that I used to determine the heat loss from the collector because I don't think a collector with such a slope would be very useful to anyone. Figure 91 shows heat loss curves for a collector with a more useful slope, 60° from the horizontal. There are four curves for four different environmental temperatures. They give the rate of heat loss from a black absorber plate which is protected by two glass cover plates spaced about an inch apart. A 10 mph-wind is blowing.

The curves in Figure 91 will also give a good estimate of the rate of heat loss from collectors with other slopes, including vertical windows or collectors with the slope of my future porch-roof collector. For instance, for an absorber temperature of 100°F and an outdoor temperature of 30°F, Figure 91 predicts a heat loss rate of 39 Btu/ft²/hr, not much different from the value of 41 Btu/ft²/hr that was found for the collector with a 23° tilt.

The rate of heat loss from a collector with only one glass cover sheet is almost double that from a collector with two sheets of glass, so a rough estimate for the heat loss rate from such a collector can be obtained by doubling the value from Figure 91. Similarly the heat loss from a collector with three glass sheets is about 2/3 that from a two-sheet collector, and so on.

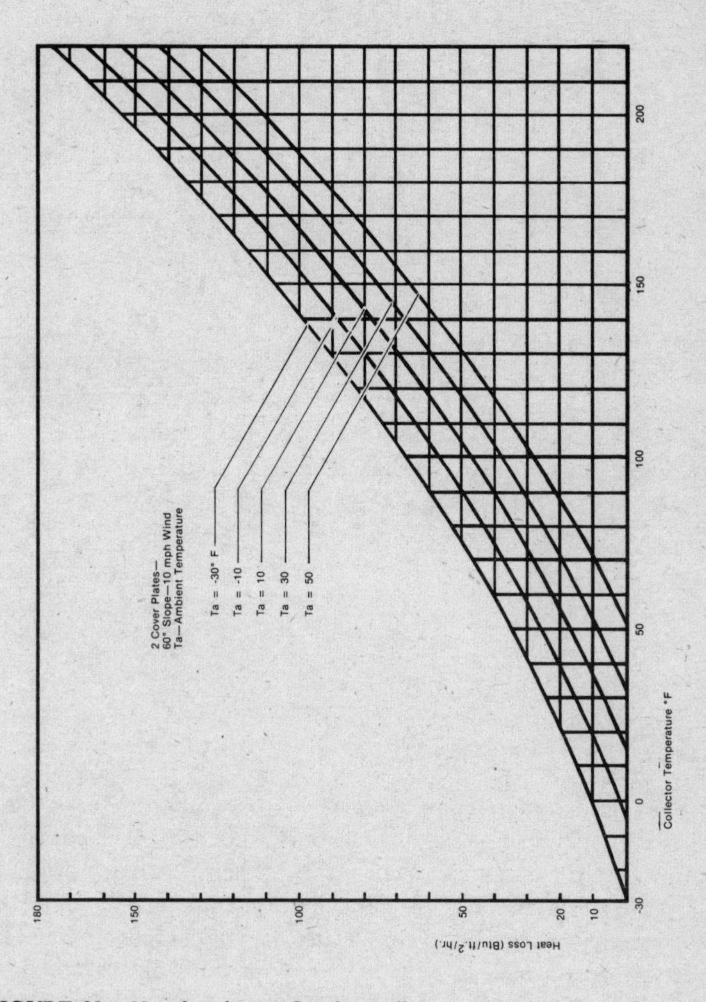

FIGURE 91. Heat loss from a flat-plate collector.

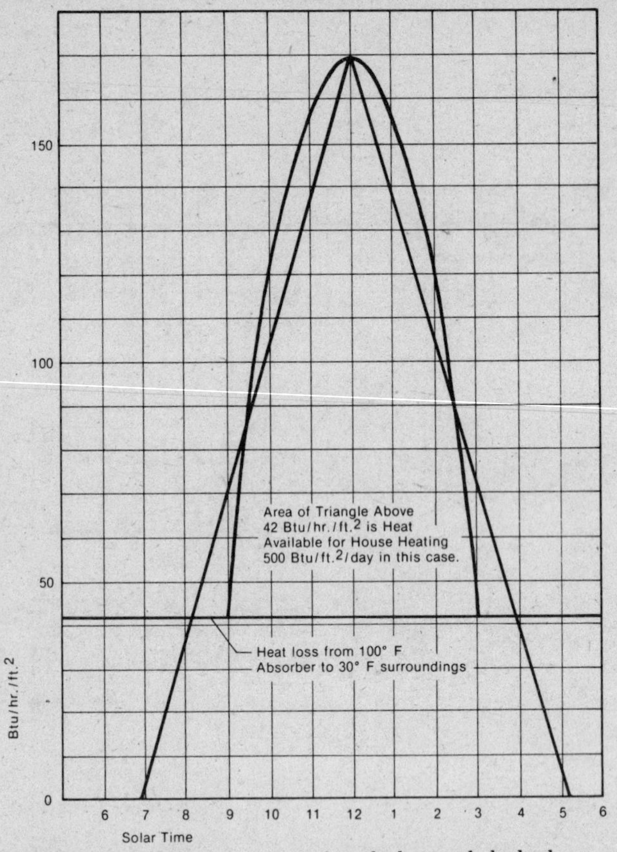

FIGURE 92. Approximate scheme for calculating whole-day heat output of a solar collector. The curved line is the actual output, the straight line, the approximate output.

At times other than solar noon the computation of the amount of direct beam solar energy that is absorbed by the collector is a lot harder than the noontime calculation, but there's an easy way to get a useful estimate of the whole-day collection. Assume that the intake of solar energy drops off linearly to zero at sunrise and sunset. The *Old Farmer's Almanac* says there are 10 hours and 15 minutes between sunrise and sunset on February 8. This means that in solar time, the sun would rise at 6:52½ A.M. and would set at 5:07½ P.M. The approximate hourly variation in the rate of solar energy absorption will be as shown by the triangle in Figure 92. The actual variation is also shown for comparison.

To simplify matters I've assumed that the outdoor temperature stays constant throughout the day so the heat loss from the collector occurs at a constant rate. It won't actually happen like this and hot air won't circulate at just the right rate to keep the absorber at 100 °F either, but it gives an estimate of the heat output.

The area of the porch roof is 42 square feet, so a collector which covered it might produce 21,000 Btu per day (42 $ft^2 \times 500$ Btu/ft^2 from Figure 92). This same amount of energy could be produced by burning a cube of wood 7.7 inches on a side. Since the 5 seconds it takes me to split this much wood is less than the time it would take me to climb the stairs and open a duct to the solar heater each day, I probably won't put a collector on the roof as long as there is waste wood to burn. All these calculations have told me that I'd be smarter to use the extra windows for my greenhouse instead.

Homemade Solar Water Heaters

Steven M. Ridenour

The easiest, most practical application of solar energy on an owner-built basis is the heating of water. Solar water heaters have been employed extensively in Israel, Australia, and Japan and were quite popular in Florida right after World War II. A carefully designed and constructed water heater can provide an adequate supply of hot water the year around. Water temperature should exceed 140°F in the spring, summer, and fall; winter water temperatures will be lower—around 120°F.

Catching the sun's rays and converting them into heat utilizes a process most of us understand. Black–painted surfaces exposed to solar energy will get hotter than any other color. A black paint which is dull or flat does not shine and thus does not lose sunlight by reflection. So if we paint a metal surface flat black and put water in contact with the back of the metal, it will be heated. This simple device is called an *absorber.*

High-temperature flat black paints can be purchased at auto supply shops—they are used for painting exhaust systems. The best

I have used consisted of a carbon black pigment and an acrylic ester resin. Such a paint will withstand 600°F indefinitely. A homemade paint of linseed oil and carbon black can be mixed by the do-it-yourselfer. Just add carbon black to the linseed oil until you reach a paint consistency. The surface will be a little shiny, but the difference in performance between this homemade paint and the auto supply kind should be slight.

Black plastics such as polyethylene, can be placed over the water instead of the painted metal. However, plastics don't stand up well under the sun's rays, and if they are used, plans must be made for a yearly change. One must also accept the environmental consequences of using plastics, a nonbiodegradable material.

Once we get the heat into the water, we keep it there with *insulation.* The heated water behind the absorber can be insulated with glass wool, straw, dry saw dust, hair felt, or polyurethane foam. Insulation between the absorber plate and the sun is accomplished with a miracle substance called *glass!* It transmits the high energy radiation from the sun which heats the water but stops the low

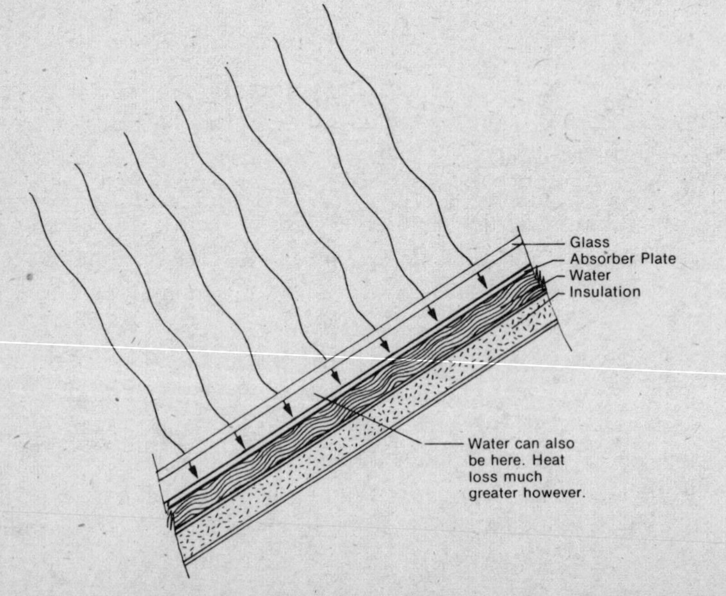

FIGURE 93. Cross-section of a solar collector.

energy infrared radiation which is radiated from the hot absorber. It also keeps air currents away from the black absorber. The substantial reduction of these two forms of heat loss makes glass an ideal insulator. Again clear plastics such as Mylar W, Tedlar, and Plexiglass can be used as covers, but their lifetimes are limited to a few years.

So there you have it: A dull black surface to catch the rays and turn them into heat; a glass cover to slow the upward heat loss; and regular insulation to stop the downward heat loss. (See Figure 93.) That is basically how it works; now for some practical applications.

How Much Hot Water?

The quantity of water needed is our first concern. Most modern U.S. homes use 25 gallons per person per day. However, if your lifestyle does not include a shower or bath every day, or an automatic dishwasher or clothes washer, then naturally you require less. I think that most people would find 12 gallons to be quite adequate. Cutting down to 4 to 6 gallons is still entirely possible for certain lifestyles.

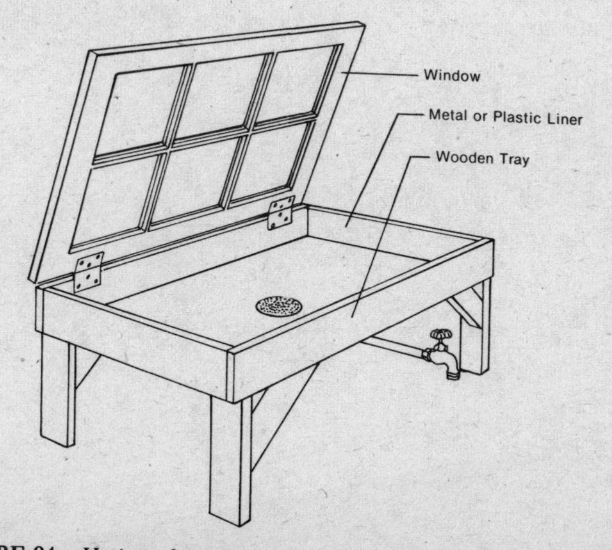

FIGURE 94. Horizontal wooden tray for solar water heater.

Simplified System: Combined Absorber And Hot Water Storage

This type of water heater is easy to construct, fairly inexpensive, and can be made portable. The disadvantages are that it must be manually filled, temperatures are usually lower than 135°F, the water will not stay warm overnight, and in fact it may freeze on cool, clear nights. Also this *batch* type of water heater does not easily blend into a piped hot water system.

WOODEN TRAY—HORIZONTAL TYPE

Line the inside of a 4-by-4-foot wooden tray with a thin galvanized iron (GI) sheet which has been soldered at the seams. Paint the inside surface flat black with a paint which is not water soluble. The wood acts as insulation, so the heavier it is, the better it will work. Black polyethylene can be used in place of the black-painted GI pan. (See Figure 94.) Fill it with 1 to 3 inches of water, depending upon your needs and the amount of sun expected. One inch of water will be about 10 gallons. In the fall and spring the sun will be able to heat only one-third as much water as in the summer. Place a window over the top of the tray. The water should reach 100–120 °F by late afternoon in the summer, spring, and fall.

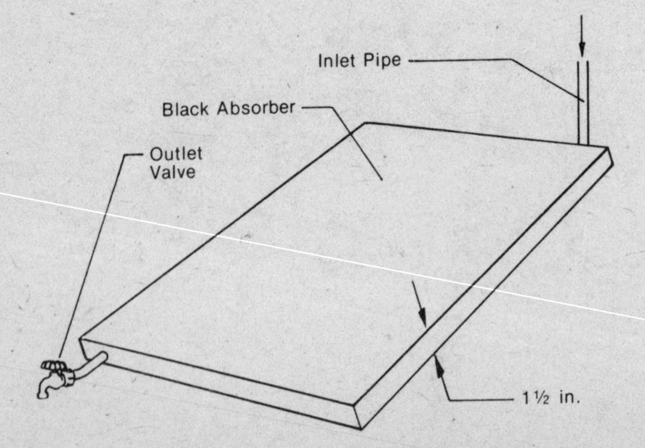

FIGURE 95. Water tank made from a sheet of thin galvanized iron.

Wooden Tray Variation—Tilted Type

In this water heater, the GI sheet depth has been reduced to 1–½ inch and a top is soldered to the pan forming a tank. Paint the top of the tank flat black. Attach an inlet pipe and an outlet pipe as shown in Figure 95. The tank need not be built if one can be purchased. Note the 1–½ inch depth; this is critical. It provides about 1 gallon of water for each square foot of area exposed to the sun. This is about optimum for useable hot water. Next, construct a box to contain the tank and place 2 inches of insulation in the bottom and sides. (See Figure 96.) Fix three legs to the box as shown. Construct the third leg so that the box tilts to a 45° angle, and cover the box with one pane of glass. Do not seal the top inlet tube; you need to allow room for expansion. A second window can be placed over the first for winter operation; separate them by ½ inch.

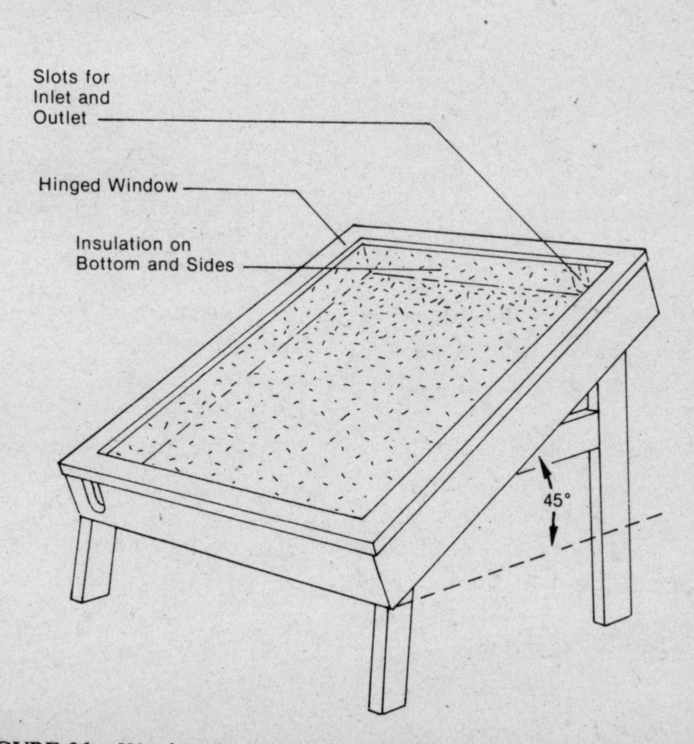

FIGURE 96. Wooden, three-legged box to hold water tank.

More Advanced System: "Fully Automatic"

This system automatically heats and stores hot water and easily adapts to plumbing systems. It is not portable, and more expense and labor are involved in its construction. As we will see later, the system can be designed to eliminate freezing. The basic parts are the collector, transfer pipes, and storage tank. (See Figure 97.)

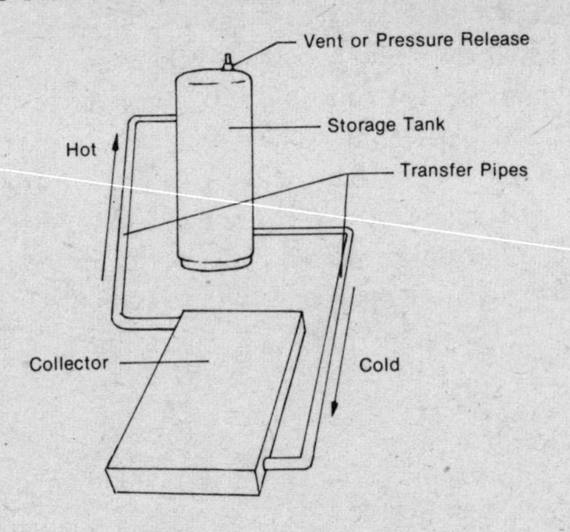

FIGURE 97. "Automatic" solar water heating system.

COLLECTOR

The size of the absorber is tied to the quantity of hot water desired. For the United States a good rule of thumb is 1 square foot of collector area for each gallon of hot water desired:

1 ft^2 Collector With 2 Glass Covers	——	1 Gallon Hot Water

For areas in the far North, such as Maine, Minnesota, Idaho, and Canada, use the following:

1½ ft^2 Collector With 2 Glass Covers	——	1 Gallon Hot Water

The above proportion is based on winter operation, so the absorber will be somewhat overdesigned for summer. The absorber should face directly south, but turning it directly southeast or southwest will affect its performance by only 20 percent. If you want hot water by noon you'll give it a southeast orientation, but if you want to wait until late afternoon you'll give it a southwest orientation. The collector should be tilted from the horizontal an angle equal to the latitude of your location plus 10°. (See Figure 98.)

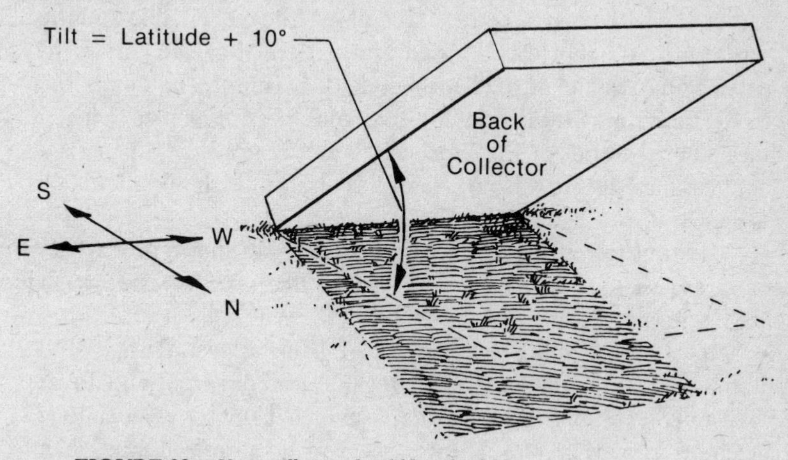

FIGURE 98. Your collector should be tilted from the horizontal at an angle equal to the latitude of your location plus 10°.

$$\text{Collector Tilt Toward South} = \text{Latitude} + 10°$$

It is comforting to know that the above relationship is not supercritical, either. The tilt can be placed at the above optimum plus or minus 10° and the collector will still work quite satisfactorily.

The actual collector for this system has many variations; most work about the same. Below I have included directions for building an absorber designed by Brace Research Institute (see Bibliography).

Building the Absorber

The following detailed instructions are based upon the use of a standard sheet of corrugated galvanized steel, 22-gauge, 8 feet long by 26 inches broad, with corrugations 3 inches apart and ¾ inch deep. If you find it necessary to use another size sheet with different corrugations, you will have to modify accordingly the dimensions set out in the sketches.

- *Step 1* Part (1). Cut the galvanized steel corrugated sheet to 26-by-88 ⅝ inches with a pair of metal shears and place to one side.
- *Step 2* Part (2). Cut the sheet of "special" flat to 26½-by-90⅝ inches. Place a piece of stiff cardboard against the end of the corrugated sheet, Part (1), and trace the shape of the corrugations on the cardboard with a soft pencil. Cut along the pencil line on the cardboard so that it can be used as a pattern. Lay the cardboard pattern on each end of the flat sheet, Part (2), in turn, and mark the corrugations on Part (2) as shown in Figure 99. With a pair of metal shears cut the ends of the "special" flat sheet, Part (2), as marked out by the cardboard pattern. The ends of Part (2) should then look as shown in the Sketch B.

    Cut out two holes, 0.84 inch in diameter, as shown on Part (2) in the sketch, to allow a ½-inch galvanized steel pipe to pass through each with a tight fit.
- *Step 3* Next take the corrugated galvanized sheet, Part (1), and attach one 9-inch length of ½-inch galvanized pipe to each end, as shown in Sketch B. First screw the 3/16-inch diameter, ¼-inch long machine screw (Part No. 5) into the ½-inch galvanized pipe, then solder the pipe to the corrugated sheet.
- *Step 4* Bend the ends of the sheet of special flat (2), at right angles, as shown in Sketch B. The bent end sections should each be 1 inch long, i.e. ¼-inch longer than the depth of the corrugations to allow for overlap when soldering. Place the corrugated sheet (1) on top of the sheet of special flat (2), slipping the ½-inch pipes into the holes. It will be necessary to cut the special flat at one end to allow the pipe to enter the hole.
- *Step 5* Bend the edges of the sheet of special flat (2) over the edge of the corrugated galvanized sheet (1), as shown in Sketch E and solder (6) as shown. To bend the edges of the flat galvanized sheet ¼-inch from the edge, clamp the sheet between two pieces of angle iron along the edge where it is to be bent and use a hammer to obtain the right-angled bend.
- *Step 6* Drill ¼-inch holes for the rivets in the valleys of the corrugations, spaced as shown in Sketch C. Place the ¼-inch galvanized rivets (3) in the holes with the heads resting on the flat galvan-

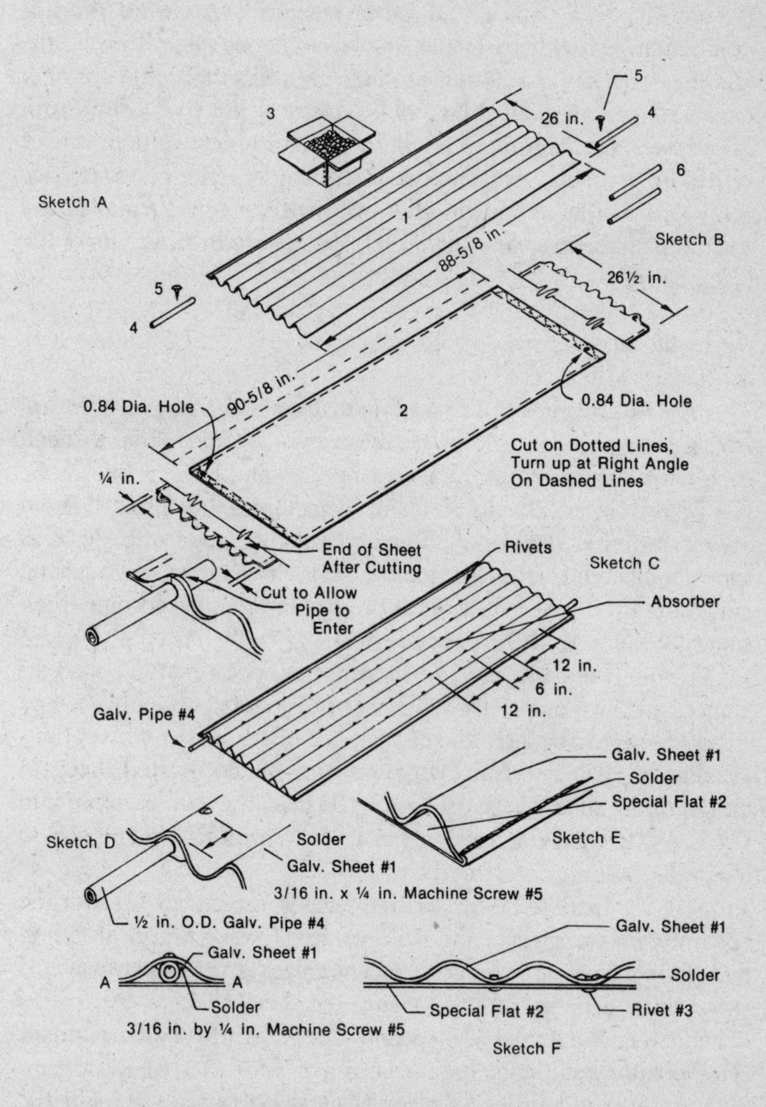

FIGURE 99. The collector absorber.

ized sheet as shown, and peen the heads of the rivets. Next solder over the peened heads of the rivets. The collector absorber is now complete and it will be necessary to test for leaks. To do this, place the absorber in a sloping position by putting it up against the side of a building or by placing a box underneath one end. Then fill the absorber with water—do not put extreme pressure on the absorber—and allow it to stand. Mark all leaks with white chalk and repair. Very slow leaks are sometimes difficult to stop—leave these, as they will most probably seal themselves with time. After all leaks have been repaired, leave the absorber filled with water to stand in the sun. Paint the corrugated side of the absorber with two coats of flat black paint.

BUILDING THE ABSORBER CASING

- *Step 7* Cut Part (7), a 3-by-8-foot sheet of flat 24-gauge galvanized steel, along one side to reduce it to a 33⅜-by-96-inch sheet. Next cut the corners off, as shown by the solid lines in Sketch G, (see Figure 101) and bend the sheet at right angles along all the lines shown dotted in the sketch. Having bent the casing into shape as shown in Sketch H, rivet the four corners with eight ¼-inch galvanized rivets (8). Place the casing aside after drilling two ½-inch drain holes in one end of the casing, as shown in Sketch J. When the unit is finally assembled in a tilted position, these holes must be located at the lower end of the casing so that any condensed moisture may drain off.
- *Step 8* Cut out Part (10), consisting of six "L" brackets and drill ¼-inch holes for the rivets in the brackets.
- *Step 9* Bond the ⅛-inch thick felt strips 1-by-1 inch (11) on to the "L" clamps.
- *Step 10* Locate the six ¼-inch holes in the casing, one in each end and two along each side as shown in Sketch J, and drill. Rivet the "L" supports (10) onto the casing, placing the flat rivet heads (12) on the outside.
- *Step 11* Draw out the coconut fiber (9) into light straw and place a 2-inch layer of fiber at the bottom of the casing for insulation. (See Sketches H and J.) (In place of the coconut fiber, 2 inches of glass wool, 2 inches of hair felt, 4 inches of packed straw, 4 inches of sawdust, or 5 inches of vermiculite may be used.) Place the

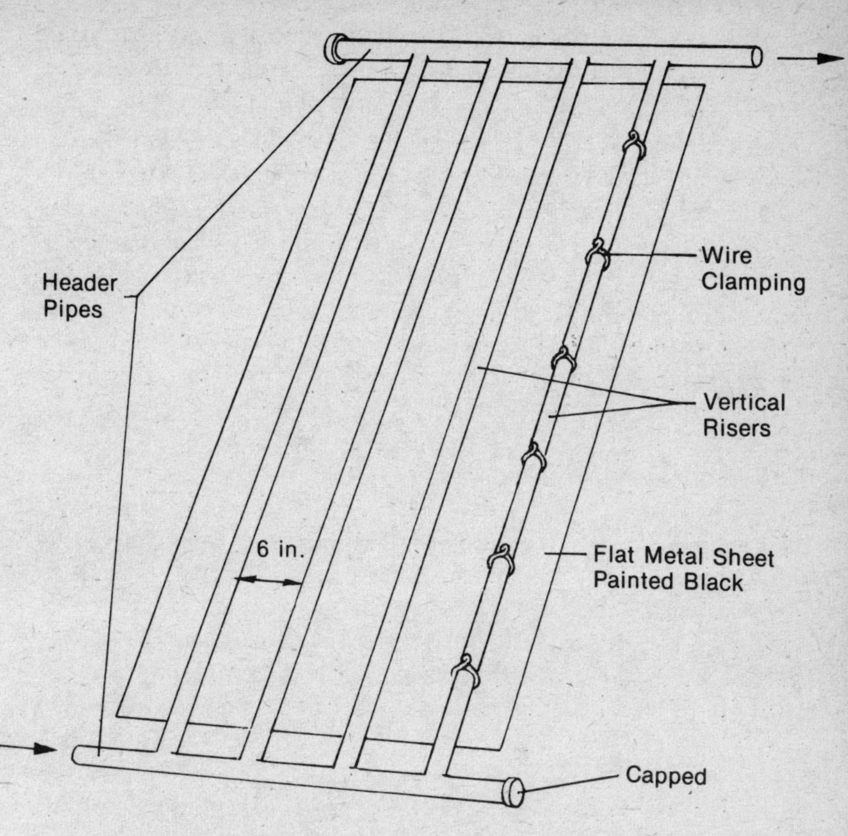

FIGURE 100. An alternative collector absorber. In many cases it will be possible to utilize flat aluminum, copper, or galvanized steel sheet for the absorber collector, as illustrated. In this case, a matrix of galvanized, or aluminum, or copper pipe or tubing can be set out on a standard flat sheet collector, normally 3-by-6 feet or 4-by-8 feet, etc. In this case the tube spacing should be approximately 6 inches, the diameter of tubing for the vertical risers should be ½ inch to ¾ inch, and the inlet and outlet header pipes should be at least 1 inch nominal piping. In the case of steel piping the vertical risers can be cut to size and holes can be drilled into the headers to receive them The ends should be welded together and the whole assembly pressure tested before attaching the matrix to the sheet. The same can be done with aluminum and copper. Under no circumstances should any dissimilar metals be used together, as this will cause galvanic corrosion. However, it is possible to use aluminum sheet and galvanized steel piping, as there is no contact between the water and the sheet. In this case reasonable thermal contact between the tubing and the sheet can be obtained by tying the piping to the sheet every few inches with galvanized wire. Obviously the closer the contact between the pipe and the sheet, the better the heat transfer, and the better the performance of the collector.

absorber in the casing, with the black painted corrugated surface upwards, resting the absorber on the "L" supports, Part (10).

- *Step 12* Make Part (13), consisting of four ½-by-¾-by 1-inch 22-gauge galvanized steel hold-down "L" clamps. Using Part (14), screw the hold-down clamps (13) into place to secure the collector absorber to the casing.

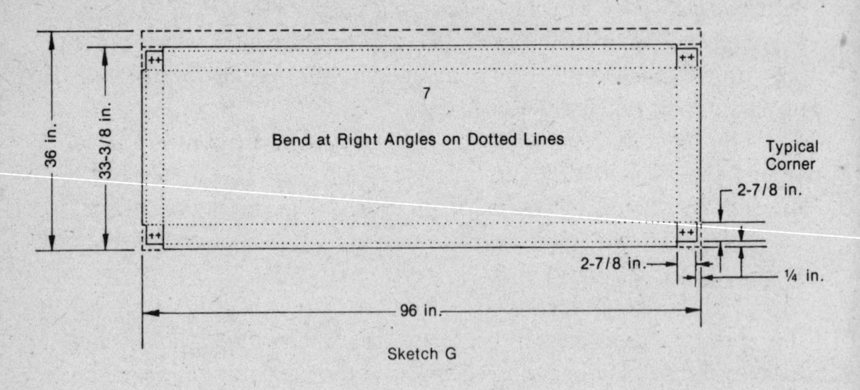

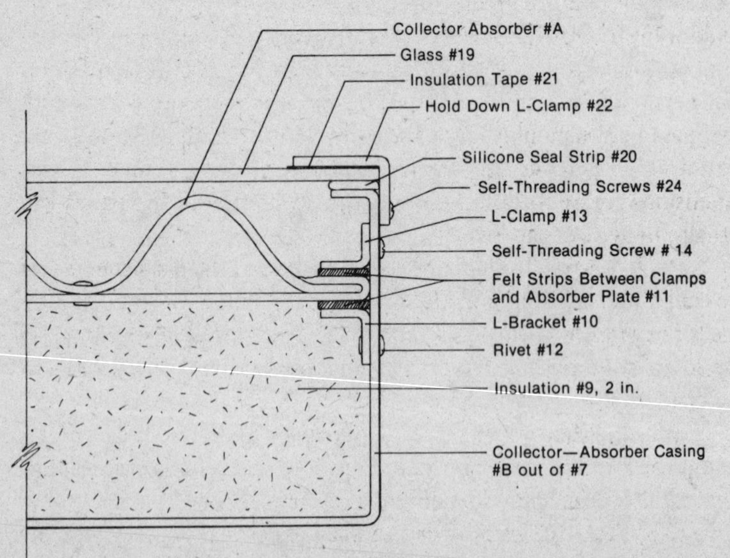

FIGURE 101. The absorber casing—Part A.

• *Step 13* Make Part (15), consisting of a galvanized steel "T" rib to support the two glass sheets at the center of the collector. The rib can be bent from a sheet of 22-gauge galvanized steel. Alternatively a pre-formed T-rib may be available commercially, e.g. the rib for the "Grecon" suspended ceiling.

• *Step 14* Rivet the rib (15) to the casing as shown in Sketch H, with two galvanized rivets (16) at each end.

• *Step 15* Stick the ¼-inch Dor-Tite or alternative sponge rubber stripping (17) onto the ¼-inch edging on the glass supporting rib (15), as shown in Sketch I.

• *Step 16* Stick ¼-inch wide sponge rubber stripping (18) all around the ¼-inch edge of the casing.

• *Step 17* Place the two sheets of glass (19) on the casing to cover the absorber, making sure that the glass rests evenly on the ¼-inch stripping (18) all round.

High-grade double-strength window glass is best for covering the absorber. The clearer the glass, the better the solar energy transfer to the absorber. The clearness of the glass can be judged by viewing it on edge. When viewed on edge, good glass will be blue and poorer-grade glass will have a greenish tinge. Greenhouse glass will have this definite green color; it is the poorest grade of glass one would want to use. Two glass covers are needed for winter hot water. Remove the top glass if summer water gets too hot. If you live in the South and have mild winters, you can use one pane year round. The glass panes should be separated by ½ to 1 inch. Also separate the lower glass from the black absorber by ½ to 1 inch. These dimensions are important—more or less separation of the panes will actually increase heat loss.

• *Step 18* Apply the silicone sealant between the glass sheets and the center and edge supporting ribs. Be sure to allow ⅛-inch on each side for expansion of the glass. The silicone sealant is very strong and should have a long life. If not available, use ordinary putty and seal the glass to the container with black electrical insulating tape.

• *Step 19* Make Part (22) 16, ¾-by-¾-by-1-inch hold-down "L" clamps, and drill an ⅛-inch diameter hole in each clamp to allow the self-threading screws to enter. Stick the ¾-inch sponge rubber strips (23) onto the glass, over the black electrical tape, where the hold-down clamps are to go (see Sketches H and J). Press the hold-

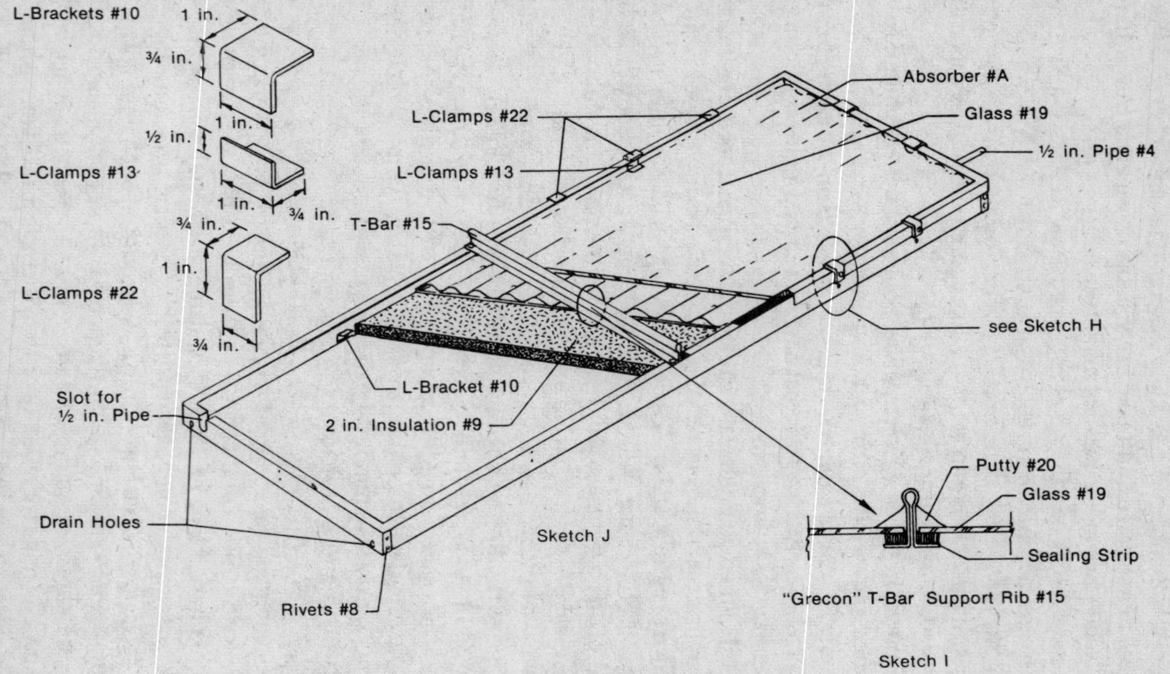

FIGURE 102. The absorber casing—Part B.

down clamps onto the ¾-inch stripping and drill an ⅛-inch hole in the casing to correspond with the ⅛-inch hole in the clamp. Use the 16, ½-inch self-threading galvanized screws (24) to screw the hold-down clamps onto the casing. The collector is now complete. Place to one side, then proceed to make the tanks.

Transfer Pipes

The transfer pipes convey the water from the absorber to the storage tank. They should be insulated with 1 to 2 inches of glass wool or the equivalent (see above), which can be taped around the pipes. The pipes themselves can be galvanized iron, copper, or even plastic; but all should be 1 inch in diameter for a collector up to 30 ft^2. (If you use plastic pipes be prepared for them to impart a peculiar taste to the water.) The pipes must be on an upward slope from the collector to the storage tank so that air cannot become trapped and stop the flow.

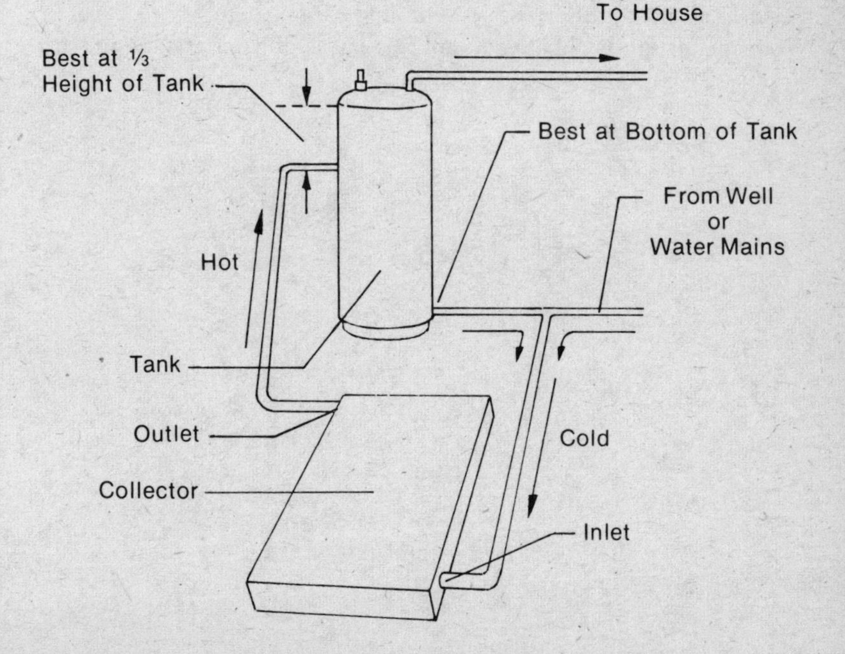

FIGURE 103. Best piping locations for a pressurized water system.

STORAGE TANK

This container should be a glass-lined insulated steel tank or a conventional insulated water heater tank. Glass or plastic lining is necessary to stop rust. A 55-gallon drum which is not so lined is not usable. The best water inlet and outlet locations are shown in Figure 103. You'll notice that the best location for the solar collector water *outlet* is about two-thirds up the side of the tank. The solar collector *inlet* should be at the very bottom of the tank. Note that the solar collector is below the storage tank—this is important and will be explained later.

Figure 103 depicts a pressurized water system. For a non-pressurized system, we must still keep the tank and collector full of water to promote natural circulation. This is accomplished with a low-level head tank. The head tank can be fabricated from a toilet tank and its associated valve. (See Figure 104.)

Important: If the system is going to be pressurized, then the tank must be able to contain the pressure. A 55-gallon drum will not work for most pressurized systems unless they are only 1 to 2 psi. If you are uncertain about your tank, hook it up to your existing system

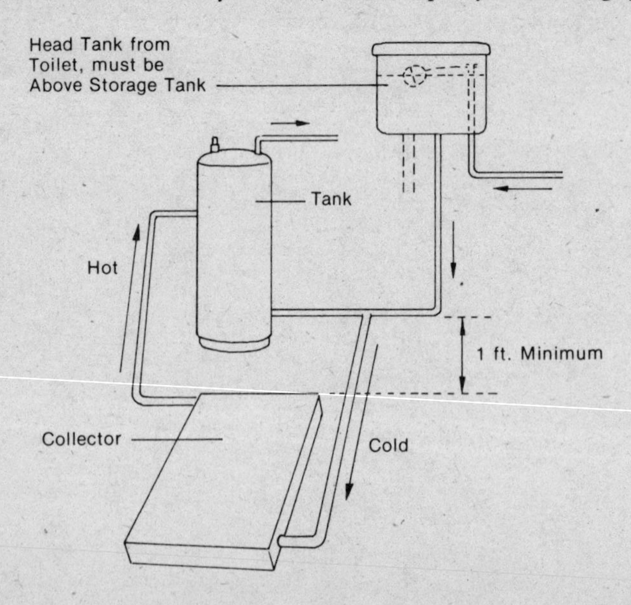

FIGURE 104. The thermosyphon principle.

and plug all holes. Slowly admit water to the tank with a valve. Let the tank sit for 24 hours under pressure, then examine it for leaks. Also check the absorber in the same manner.

The Thermosyphon Principle

If the system is properly designed and constructed, the thermosyphon principle eliminates the need for a circulating pump between the absorber and the storage tank. Just as warm air rises over cool air, warm water rises over colder water. (See Figure 104.) The water in the absorber absorbs heat produced by the sun striking the flat black surface, and as it heats, its density decreases. It rises to the top of the absorber and goes through the transfer pipe to the storage tank. Actually it is being pushed up by the heavier water in the cold water inlet pipe to the absorber. Cold water in the bottom of the tank flows to the bottom of the collector, is heated, and rises to the top of the tank. Thus a flow is established and continues until the water fails to gain energy (heat) from the sun.

At night we see that the tank is above the absorber and contains the hot water from the day's work. Since the hot water cannot seek a higher level, it is stable in the tank until needed. (See Figure 104.) However, if mistakenly the top of the collector is somewhat above the bottom of the storage tank, as illustrated in Figure 105, then the thermosyphon can operate in reverse at night and actually chill the water below the lowest outdoor temperature. Important: Always place the storage tank at least 1 foot above the collector; 2 to 5 feet is even better.

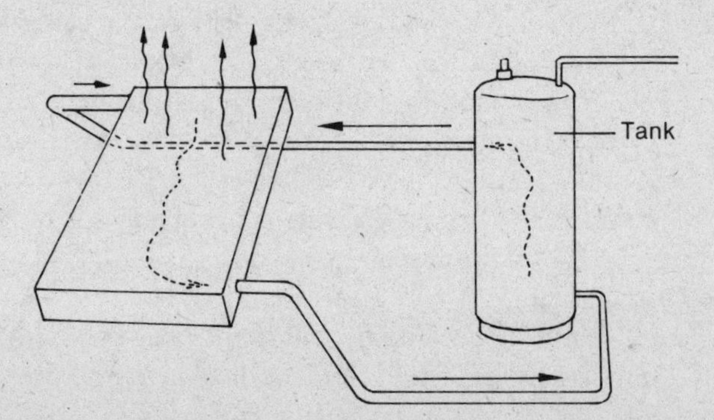

FIGURE 105. Night operation, improperly placed collector.

THE SEPARATE COLLECTOR LOOP

The best way to eliminate freezing of the absorber plate in winter is to add antifreeze to the water which flows through it. This is best accomplished by a separate heat transfer loop for the collector system. (See Figure 106.) Be sure to insulate all pipes, especially the heat exchanger. The exchanger can be made by wrapping ¾ inch copper-tubing around the storage tank. About 20 feet of the tubing wrapped around the lower two-thirds of the tank will be sufficient. If the tank is covered with insulation, then this must be removed so that the copper tubing is in contact with the tank. Insulate the

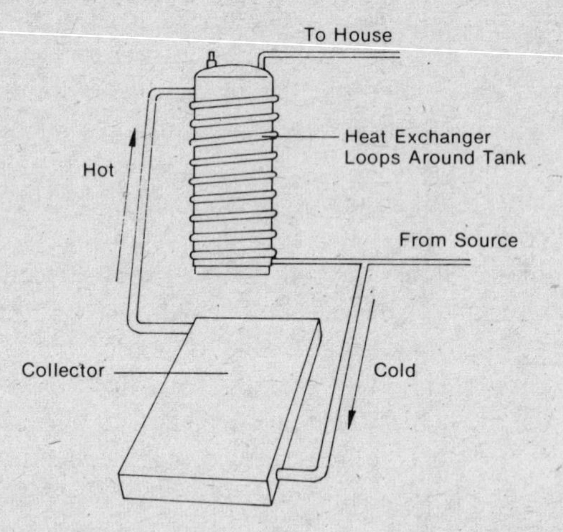

FIGURE 106. Solar hot water system with heat exchanger.

tube-wrapped tank with 2 inches of fiberglass or the equivalent.

Another nice advantage of having this loop is that you can pressurize the storage tank without pressurizing the absorber plate, and this can make the design and construction of the absorber much simpler. Two problems are solved in one.

You may find that you will want a conventional gas or electric water heater to back up your solar heater during cloudy cool weather or when you need an unusually great amount of hot water. Auxiliary water heaters can be hooked up to your home-built unit fairly easily since the automatic solar system I've described here is so simply laid out.

Design Sheet and Checklist

Design

1. Quantity of hot water desired per person per day (use maximum) ______ Gal.
2. Total number of people consuming hot water ______
3. Size of storage needed
   $= 1 \times 2 + 20\%[1) \times 2)] =$ ______ Gal.
   Note: For system with no auxiliary water heater, multiply above by 2 to allow for one day's reserve ______ Gal.
4. Size of collector needed
   $= 3$ ______Gal. $\times$ 1 ft²/Gal. $=$ ______ Ft.²
   Note: Areas in the far North use $= 3$ ______Gal. $\times$ 1½ ft²/Gal. $=$ ______ Ft.²

Checklist

1. Two glass panes or plastic covers?
2. Storage tank above collector?
3. Storage tank vented? Insulated?
4. Transfer pipes insulated?
5. Collector orientation
   a) Due South
   b) Southeast or Southwest
   c) Tilt = Latitude + 10°

TABLE 25. Materials List

PART NO.	NO. OF PARTS	MATERIAL	SIZE
A - The Absorber			
1	1	corrugated galvanized steel sheet	22–gauge, 8 ft. × 26 in.
2	1	"special" flat galvanized steel sheet	22–gauge, 8 ft. × 36 in.
3	28	galvanized steel rivets	¼ in. in diameter, approx. $^5/_{16}$ in. long
4	2	galvanized steel water pipe	½ in. in diameter, 9 in. long
5	2	m.s. machine screw	$^3/_{16}$ in. in diameter, ¼ in. long
6	2	sticks of solder	
B - The Absorber Casing			
7	1	"special" flat galvanized steel sheet	24–gauge, 8 ft. × 3 ft.
8	8	galvanized rivets for ends of casing	¼ in. in diameter, approx. $^5/_{16}$ in. long
9	–	coconut fiber or equivalent insulation	20 lbs.
10	6	22–gauge galvanized steel sheet	1 in. × 1 in. × ¾ in., supporting "L" brackets
11	6	felt strips or suitable insulation	1 in. × 1 in. × ⅛ in. thick
12	6	galvanized rivets for part (10)	¼ in. in diameter
13	4	22–gauge galvanized steel sheet	1 in. × ¼ in. × ½ in., hold-down "L" clamps
14	4	galvanized steel self-threading screws for part (13)	⅛ in. in diameter × ½ in. long
15	1	22–gauge galvanized steel sheet	27⅛ in. × 2 ½ in., to make glass-support rib
16	4	galvanized rivets for part (15)	¼ in. in diameter, approx. $^5/_{16}$ in. long
17	2	sponge rubber strip (e.g. "Dor-Tite")	¼ in. × ⅛ in. × 17¾ in. long
18	1	sponge rubber strip (e.g. "Dor-Tite")	¼ in. × ⅛ in. × 22 ft. long
19	2	window glass	27¾ in. × 44¾ in. × ⅛ in. thick
20	1	silicone type sealant (or equivalent)	12 oz. cartridge
21	1	black plastic electrical insulating tape	one roll, 1 in. wide (or nearest)
22	16	22–gauge galvanized steel sheet	1 in. × ¾ in. × ¼ in., hold-down "L" clamps
23	12	sponge rubber strip (e.g. "Dor-Tite")	¼ in. × ⅛ in. × ¼ in. long
24	16	galvanized steel self-threading screws	⅛ in. in diameter × ½ in. long

APPENDIX I. COMBINING ALTERNATE ENERGY SYSTEMS

Donald Marier, Ronald Weintraub, and Sandra Fulton Eccli

In nearly all American homes, sun, wind, water, human wastes, and organic garbage are enemies, to be kept out or disposed of. In summer, the sun pours in through unshaded windows, turning the house into a Turkish bath. In winter, wind whistles through cracks around loosely fitting windows and doors. In spring, rainwater deteriorates roof shingles and floods the basement. Human wastes and garbage require expensive, difficult-to-maintain septic tanks or sewage treatment plants.

Yet these elements need not be enemies. The technology now exists to put all of them to work. Solar power, wind energy, water energy, and fuels made from organic matter (animal and plant wastes, including human fecal matter) will be the best future sources of power for all of us—our best friends.

In this age of specialization, we are accustomed to looking for the "single" solution—the one cure-all—to solve our problems, and most Americans tend to look for such a cure-all to end our energy woes. But it should be understood that any single source of energy —whether it be sun, wind, or what have you—will not be sufficient to supply the complete energy needs of a homestead, farm, or small community. We must think in terms of combining these sources, to work in harmony with each other and with the natural environment.

Why should one source—say, solar power—not be adequate to supply all the power needs of a community? For one thing, the supply of most of these new resources is intermittent. And each energy source comes to us in a different form. Solar energy arrives as heat, and it is most efficient to use it as heat. Wind and water come in mechanical form. Fuels from organic matter (methane gas and alcohol are two of these) are generally more portable and versatile, but not that plentiful. So, our combined energy system will have to consist of a diverse and harmonious matrix of the various energy sources, their storage, interconnections, final energy use, and, where possible, recycling back into the system.

We must now talk about "energy conversion"—a term many readers may not be familiar with. An "energy conversion" is when a form of energy, say, heat, changes into another form of energy (sometimes more useful), such as electricity. One of the aims of using multiple energy sources is to minimize the number of energy conversions which take place, because each conversion usually involves some loss of efficiency.

Consider how five basic forms of available energy interact with one another. The interactions, pictorialized in Figure 107, involve solar, chemical, mechanical, electrical, and heat energy. Whenever possible, it is best to use direct conversion. "Direct conversion" of energy results when there is only one conversion necessary; that is, only one line is shown in Figure 107. Familiar examples are the conversion of solar energy to stored chemical energy in plants, through photosynthesis; or the conversion of that chemical energy into heat, by means of combustion of plant matter.

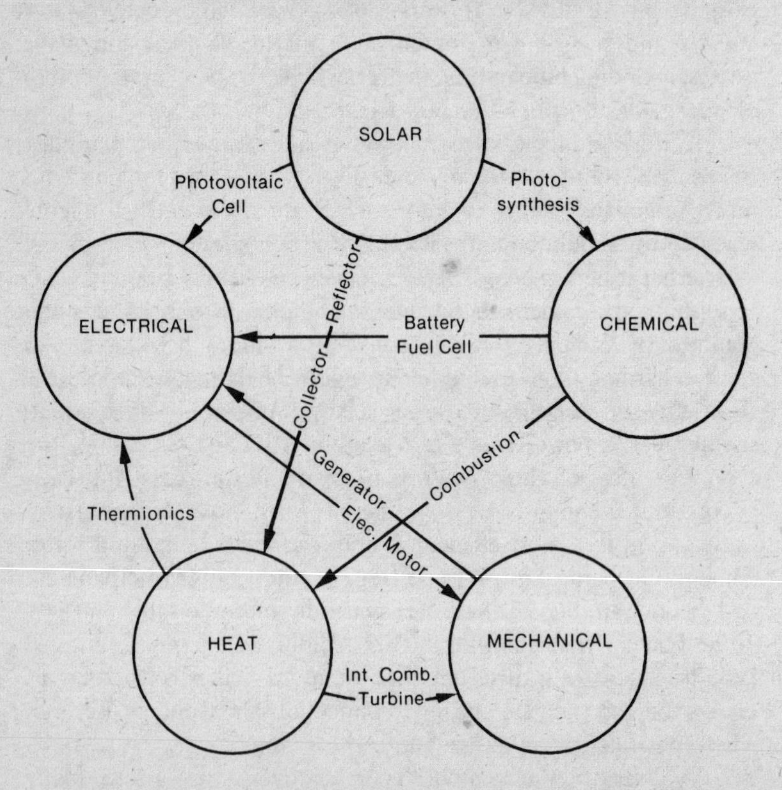

FIGURE 107. The interaction of five basic forms of energy.

In general, then, we can say that the greater the number of conversions between the basic forms of energy the greater the waste and the higher the chance for a decrease in efficiency of a process.

The sources from which energy is obtained are often called "conversion sources." For our purposes, we are dealing with the sun, wind, water, and organic material. The most important thing to remember is that (unlike fossil fuels) the supply is free and will last indefinitely.

Free or not, we had better plan to use our sources efficiently, or the cost in dollars will be astronomical. This includes not only minimizing energy conversions, but also matching time patterns to make use of whatever source is available, as in Table 26. (For example, solar energy is most efficiently used on sunny days.) However, one-to-one matching will not always be possible. Heat may be needed during the nighttime hours or on cloudy days, and that is why plans for a solar energy collection system must include plans for storage. The closer the matching of "supply time" and "demand time," the less storage will be needed, and the more efficient and less costly the system will be. When storage is needed, a system which itself minimizes energy conversions, such as the pumping of water (which can be done any time there is an excess of energy) is definitely favored over the use of a storage device such as a battery, which essentially stands outside the combined system of sun, wind, water, and organic fuels.

A small scale energy system, then, must be designed so that no available incoming energy is wasted. All of it must be either utilized on the spot, or stored for later usage.[1]

Let's take a look at the various energy systems we are discussing, in terms of their efficient utilization and including possible methods of storing the energy for future use. Figure 108 will be useful in this discussion.

Wind Power

Where water power is not available and the wind averages 10 miles per hour or greater, wind power will be used for almost all non-heat loads. An efficient method of using wind energy is the

TABLE 26. Loads, Energy Sources, and Timing Requirements of an Agricultural Community Using a Combined Alternative Energy Power System

RANDOM (DAY OR NIGHT)	DAY (NO PRECISE TIME)	DAY (PRECISE TIME)	NIGHT (PRECISE TIME)	ENERGY SOURCE: WIND	SOLAR	METHANE	WATER	NOTES
Domestic water				WD		M	W	M=supplement W=hydro ram
Water distillation					S			
Irrigation				WD		M		M=supplement
Water heating, steam-raising				WD	S	M		WD & M= supplements
	Threshing			WD		M	W	M=supplement
	Grinding, food mixing, fodder chopping			WD			W	
	Small industrial or agricultural power			WD		M	W	In combination
		Cooking		WD	S	M		In combination
		Fans, refrigerator air conditioning		WD	S			In combination
			Lighting	WD		M	W	With battery storage
			Radio & small domestic power	WD		M	W	W=w/battery storage

generation of electricity directly from the propeller shaft and storing it in three possible ways: (1) in storage batteries; (2) as hydrogen and oxygen electrolyzed from water; or (3) to run a compressor from the propeller shaft and store the energy as compressed air. All of the homestead or community's mechanical and electrical needs, then, would be handled solely from the stored supply. This method is simple because all energy input flows through one terminal. However, the cost of the very large amounts of storage needed can offset the advantage of simplicity. Golding estimates that the energy supplied to a farm via a battery by this method will cost about six times more than if supplied by only random power—that is, only when the wind is blowing.[2]

Wind is best used as mechanical or electrical energy rather than heat energy. Also, since wind is an intermittent source (perhaps several days may go by when the air is still), the capacity of a wind storage system can be made smaller by using an auxiliary source of mechanical power to supplement the wind energy on calm days. Conceivably this could be supplied by an auxiliary fuel such as methane, alcohol, or wood.

Solar Energy

Solar energy is most efficiently utilized directly as heat. Although research is being done on the utilization of solar energy in solar cells, these are presently too costly for the average person. Solar powered mechanical devices are relatively inefficient at this time and would probably be limited to very sunny climates. Therefore, excess solar energy will almost always be transferred to heat-storage units for further use. There are various heat-storage units for storing the sun's energy, but the ones in most common use are water, rocks or gravel, solid blocks, and salts such as Glauber salt. All of these store solar energy either by raising their temperature or by changing their physical state. They return the solar energy in the form of heat. Such storage allows for the use of solar energy at night and on cloudy as well as sunny days.

In very sunny regions, other devices for using solar energy, like thermoelectric and photovoltaic devices and hot air Stirling engines, are theoretically available—at least for further experimentation.[3]

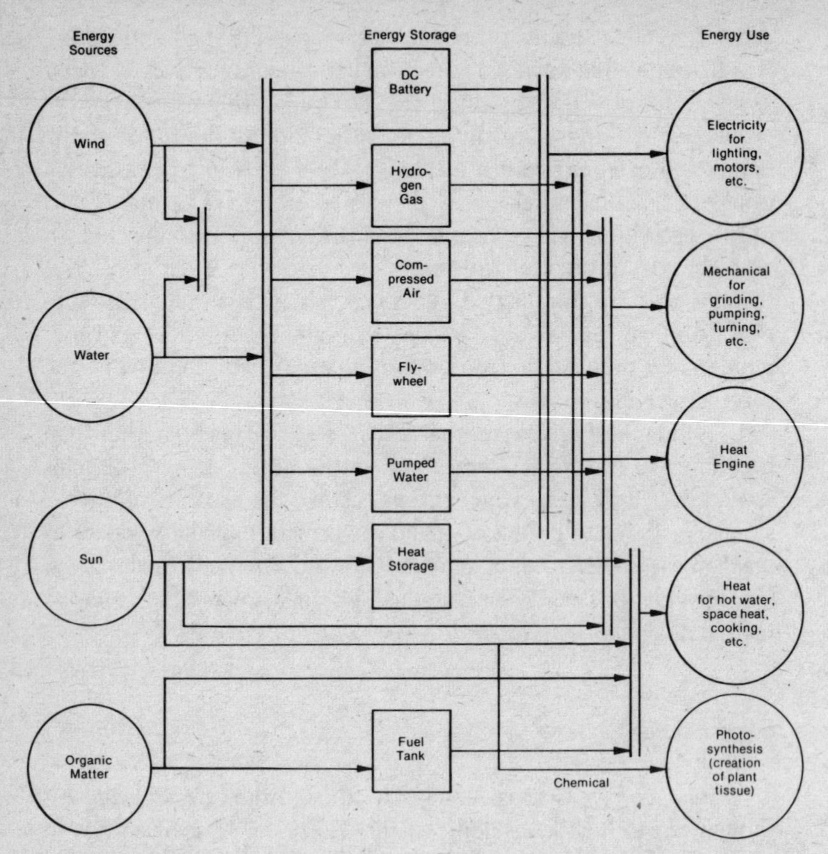

FIGURE 108. Combined energy storage and use in integrated systems.

Water Power

The best water power sites are those where a waterfall or a dam already exists. An overshot wheel can be set up at such sites to transfer the energy of the water into mechanical work or electric power. Overshot wheels are simple and useful for a power development of up to about one horsepower. Water can also be used to pump itself through a simple device known as a hydraulic ram.[4] Once pumped to an uphill storage site, it can be stored in the form of potential energy for later use in irrigation, etc.

Harnessing a small stream for water power can be quite advantageous.[5] Water power, however, while not an intermittent source in the same sense as solar power with its day-night fluctuations, or wind power with its random ones, can have seasonal fluctuations. Some streams that run swift and deep in the spring are almost dry by August. This certainly must be taken into account in designing your total energy system.

Organic Resources

Organic sources of fuel—chiefly methane, alcohol, and wood gas—are the most versatile of the energy sources the average community is likely to be able to develop. Their use depends upon their availability in a given location, but it is not restricted to a particular time of day or year.

If fresh organic material is left to digest in an airtight container, anaerobic bacteria break it down into simple, more stable chemical substances. The final products include methane gas and an excellent fertilizer as a bonus.

Alcohol is obtained by fermentation and distillation of various organic substances. As a fuel, it has enormous potential. Wood—now used inefficiently and wastefully by being partially carbonized in open fireplaces—can be utilized much more efficiently by being converted into wood gas in a modern wood burner. Wood gas can produce three times the heat from a given log that the old-fashioned fireplace or Franklin stove was able to produce.[6]

How to Figure Out Your Energy Needs

These, then, are the basic systems with which you must familiarize yourself. Next you need to know your own basic energy requirements—whether they are as small as those of a single person or small family, or as large as those of a community. Naturally, the energy needs of a farm or community vary greatly, depending upon social organization, personal lifestyle, climate, and local geography. However, for an estimate of energy needs, Table 27 can be useful. Timing requirements are taken into consideration so you can best

TABLE 27. Estimates of Local Energy Requirements for a Small Community of One Family (Four People)[7]

ENERGY USE	ANNUAL ENERGY (KWH)*	TIMING REQUIREMENTS	PROBABLE ENERGY SOURCE	BASIS OF ESTIMATION
Irrigation	300/acre-ft	Random	W, Wa, S, Or	See Example 2
Grinding, food mixing & fodder chopping	50/family	Random	W, Wa, Or	Average of 1 to 2 tons/ family/yr, including animals
Domestic water Pumping	50/family	Random	W, Wa, Or	150 gal/day. See Example 1
Lighting and small power	850/family	Night & day, precise	W, Wa, Or	5–100 watt lights used 4 hrs/day, ⅓ kwh/day misc. items—power tools, stereo, appliances, etc.
Misc. small power	100/family	Night & day, precise	W, Wa, Or	5 kw total power
Refrigeration	200/family	Random	W, Wa, S, Or	½ kw unit run 1 hr/day
Space heating**	4100–15,000	Day & night, precise	S, Or	7.3–10 kw average heat requirements. 60–750 kwh heat storage based on a survey of existing solar houses
Water heating	3500/family	Random	S, Or, W	50 gal/day
Cooking	400/family	Day, precise	Or, S	1–1.5 kw stove used 1 hr/day
Cultivation	5600/25 hp/ 75 hrs	Variable	Or, W, S	25-hp tractor used 75 hrs/yr
Transportation	7500/vehicle of 25 hp/100 hrs use	Variable	Or, W	25 hp vehicle used 100 hrs/yr See Example 3

OR = Organic fuels; W = Wind; S = Solar; Wa = Water
*1 kwh = 3413 Btu
** See also Table 29.

balance energy input against output and also estimate the amount of storage that will be needed.

Using Table 27, persons who can do simple arithmetic can estimate their total yearly energy requirements as expressed in kilowatt hours (kwh). Here are three examples:[8]

Example 1—Pumping Domestic Water: Assume a typical family uses 150 gallons per day. Assume the water pump can move 500 gallons/hour (1.1 feet/minute). Then the pump will run 3/10 hours per day or 110 hours per year. Assume the pump must lift the water against an effective "head" of 200 feet. Then, from the equation P(power) = 1.89 × .001 × F(flow) × H("head" or vertical distance to be pumped) the power required is: P = .32 kw, or (assuming 70 percent pump efficiency) .46 kw. The total energy per year is then: .46 × 110 hours = 50 kwh.

Example 2—Pumping Irrigation Water: Find out the depth from which your water is pumped and how many gallons of water your pump can move in one hour. Assume irrigation water is pumped from 200 feet below the surface and, using the same argument as Example 1: At 500 gallon/hour the pump will have to run a total of 650 hours consuming 0.46 kw of power = 300 kwh/acre-foot (1 acre-foot is about 325,000 gallons).

Example 3—Power for Transportation: Assume an economy engine of 25 horsepower and a 2.98 kw input for each 1 hp of developed power.[9] If this vehicle is used for general transportation 200 hours a year, the annual energy need would be 4,900 kwh. At an average speed of 40 mph, 200 hours of use per year would be the equivalent of driving 8,000 miles. Assuming that future mass transit for long-distance travel will cut the vehicle use in half, the annual energy use would then be 7,450 kwh.

Calculating the Available Energy in Your Area

The above were your energy needs. Now you need to figure out how much energy you can draw upon to fill those needs. There are typical amounts of available energy, which are given in Table 28. They may, however, not be typical for your area, and you should fill in local values. The reader can draw up tables similar to Tables 27 and 28 for any specific situation as an aid in estimating whether energy needs can be supplied by local available sources. Please remember that the figures you put into your tables depend upon your need and your local energy resources.

To calculate the energy potentially available in your area, you must first collect "raw data"—that is, the energy available before conversion.

Solar Data: Data or solar radiation for different parts of the U.S. can be obtained from the U.S. Weather Service,[11] the American Society of Heating and Air Conditioning Engineers' Annual Guide,[12] or from the Climatic Atlas.[13] Information is usually given in Btu/feet^2/day or in gram calories per square centimeter (or Langleys) per day. For a correctly tilted solar collector, the values may run as much as 15 percent higher in the winter months.

Wind Data: Average wind speed in many parts of the country can be gotten from the Weather Service, the Climatic Atlas, or from local airports.

Water Flow: Must be computed at the site. Methods are fairly simple. (See section on measuring flow rate in the chapter, **"Small Water Power Sites."**)

Methane: Calculate the available digestible wastes per year per animal, plant, and human. Figures can be obtained from Table 16 (See **"Methane Gas Digesters for Fuel and Fertilizer"**). Multiply the number of pounds per day by the number of animals, then multiply by 365 to get the number of pounds per year, your total.

Energy Storage: The actual amount of storage needed for each form of energy is very difficult to estimate and is not, therefore, included in any of the tables. But it is useful to try to estimate it while estimating available energy resources because you will be contacting many of the same agencies for data. A great deal depends upon the maximum number of windless or sunless days that can be expected. The Weather Service will sometimes have this information, or at least have records of wind speeds and cloud cover from which the greatest probable number of windless and sunless days can be computed. Other sources would include local newspapers and airports.

Designing a Combined Energy System for Your Needs

Three basic factors are needed in order for you to design an optimum system for your needs. You must know your energy requirements (Table 27) and your available energy (Table 28), and of course your financial resources, or those of your community. The amount of money needed to build a combined energy system is subjective and depends on such factors as: what skills are available; what local materials are available; future development of alternate business and economies which might include such practices as bartering; and certainly the climate of your area.

When you know all three factors, you are ready to start designing your combined energy system. Let's look at a typical (hypothetical) situation—a semi-farming community in the southwest U.S.[14]

The Southwest (New Mexico) community might include 20 people. The climate is semi-arid, and of the available land, 40 acres are used for grazing, 15 for irrigated agriculture, and 15 for housing and living. Domestic animals include 20 goats, 40 chickens, and two horses. Excess dairy products are sold or traded for other needs and for available items such as propane gas. The buildings include a small barn, a chicken house, a greenhouse (self-heated, partially underground), and 4500 square feet of personal dwelling floor area. Refrigeration, auxiliary space and water heaters, washers, tools, vehicles, farm equipment, and most appliances are shared on a cooperative basis for greater efficiency. There are two transport vehicles—a truck and a car—and one 25-hp tractor.

Energy Needs

a. Irrigation: Water needs for various crops can total from 1.0 to 3.4 acre-feet per acre per year.[15] Water losses in hot dry climates can be about 30 percent. If we assume a water need of 2 acre-feet/acre/year and a 30 percent loss, 2.9 acre-feet will actually be required. Natural rainfall will contribute only about 0.7 acre-feet/-year in this area, leaving 2.2 acre-feet/acre to be supplied by irrigation.

b. Grinding, fodder chopping, and food mixing: This assumes the community will have the following grain requirements:

Humans	300 lbs/person/yr	= 6,000 lbs/yr
Horses (medium, work)	3,000 lbs/horse/yr	= 6,000 lbs/yr
Goats	548 lbs/goat/yr	= 10,095 lbs/yr
		22,095 lbs/yr

A grain composed of corn and oats requires approximately 18 kwh/ton for mixing, or a total of 21 kwh/ton. Therefore, the energy needed in New Mexico would be 231 kwh.

c. Domestic water pumping: For a 200-foot well, 250 kwh would be needed for 20 people.

d. Lighting and small power: Due to considerable savings of energy per person in a communal-living situation, the estimated energy needed by the 20-person New Mexico settlement is assumed to be only three times the value of 850 kwh for one family, or 2,500 kwh.

e. Miscellaneous small power: Again we assume the saving of energy in a community situation (three times the energy required per family) =300 kwh.

f. Refrigeration: 1½ kwh/day × 3 (for 20 people) = 600 kwh.

g. Space heating: The annual heat load of a building in Albuquerque is approximately 124 million Btu/year. For a dwelling with a 1500-ft^2 floor area, 4,500 ft^2 would require 369 million Btu's or 108,000 kwh.

h. Water heating: Annual water heating load in Albuquerque area is approximately 918 therms (26,900 kwh) for a 4,500-ft^2 dwelling, enough to heat 140,000 gallons of water by 80°F.

i. Cooking: A number of sources consider a stove to be about a 1-kw power unit. A meal for 20 persons would require only about three times the energy for four people, thus annual energy need would be about 1,200 kwh.

j. Cultivation: A 25-hp tractor used 75 hours/year, along with two horses = 5,600 kwh.

k. Transportation: Assuming a truck and a car = 15,000 kwh.

AVAILABLE ENERGY

a. Water power: None.

b. Solar: The average daily solar energy received at Albuquerque is approximately 1900 Btu/ft^2/day. Assuming there is a 50 percent efficient solar collector, the useful energy would be 950 Btu/ft^2/day. Converting this to kwh/ft^2, or 950 $\times$ 2.93 $\times$ 10^{-4} kwh/Btu $\times$ 365 days/year = 101.6 kwh/ft^2/year. (Use Table 29.)

c. Wind: Assume a 10 mph average, and a 15-foot diameter windmill. Power generated = .445 kw $\times$ 8760 hours/year = 3898 kwh/year. (Use Table 30.)

d. Methane: Available digestible wastes per year per animal = 18,026 pounds/year. Assuming that all manures could be collected and that 1 pound of digestible matter yields 5 ft^3 of gas, then 90,130 ft^3 of gas would be produced. At a heat value of 700 Btu/ft^3, this

TABLE 28. Typical Amounts of Local Energy Available for a Rural Settlement (from Basic Conversion Sources)

SOURCE	ENERGY AVAILABLE	BASIS OF ESTIMATE
Solar	65 kwh/ft^2/yr	Assume location near Madison, WI. From Table #29, average daily solar energy received there is 1218 Btu/ft^2/day. Assuming a 50% efficient solar collector, useful energy would be 609 Btu/ft^2/day. Converting this to kwh/ft^2, or 609 Btu/ft^2/day $\times$ 2.93 $\times$ 10^{-4} kwh/ft^2/yr.
Wind	4,000 kwh/yr	Assume a 15-ft diameter wind generator where the average wind is 10 mph. From Table #30, power generated = .445 kw, or .445 kw $\times$ 8760 hrs/yr = 3898 kwh/yr.
Water	9,800 kwh/yr	Assume a water power site has a flow of 100 ft^3/min and a head of 8 ft. Use equation: P(power) = 1.89 $\times$ .001 $\times$ F(flow) $\times$ H(head, or vertical distance the water drops). Then, P = 1.5 hp, or 1.5 hp $\times$ .745 kwh $\times$ 8760 hr/yr = 9789 kwh/yr.
Alcohol	1,500 kwh/acre	Alcohol yield from potatoes = 178 gal/acre. Assume distillery is able to obtain 60 gal/acre; or, 60 gal/acre $\times$ 84,000 Btu/gal $\times$ 2.94 $\times$ 10^{-4} kwh/Btu = 1476 kwh/acre
Wood	3,000 kwh/acre/yr	Assuming a wood yield of 1 ton/acre/yr and 7,000 Btu per lb of wood, 14 million Btu/acre/yr would be available.[10] With a 75% efficient wood stove & automatic damper, the amount of energy in kwh would be: .75 $\times$ 14 $\times$ 10^6 Btu/acre/yr $\times$ 2.94 $\times$ 10^{-4} kwh/Btu = 3,076 kwh/acre/yr
Methane	3,520 kwh net	Assume 1 cow at 8 lbs/day = 2,920 lbs; 1 hog at 1.3 lbs/day = 475 lbs; 4 humans at .25 lbs/day = 365 lbs; and 2,000 lbs plant wastes = annual yield of 5,760 lbs. 1 lb of digestible matter gives 5 ft^3 of gas, then above wastes would produce 18,500 ft^3 of methane gas, or 4,400 kwh gross. If about 20% of gas is used to keep the digester at 95°F, then 3,520 kwh would be available for other uses.

TABLE 29. Daily Averages (Btu/ft^2/day) of Solar Energy Received on a Horizontal Surface By Months (From Heating and Ventilating Reference Data, 1954)

CITY	JAN	FEB	MAR	APR	MAY	JUN	JUL	AUG	SEP	OCT	NOV	DEC	ANNUAL
Santa Maria, CA	1070	1380	1882	2251	2506	2399	2428	2369	1945	1594	1114	867	1817
Grand Lake, CO	790	1144	1624	2030	2177	2362	2236	1989	1720	1328	863	613	1373
Miami, FL	1100	1284	1535	1756	1852	1771	1749	1716	1528	1351	1232	1085	1497
Griffin, GA	1063	1107	1181	2103	2288	2273	2170	2066	1546	1358	1044	738	1578
Twin Falls, ID	613	827	1269	1705	2184	2303	2280	1985	1646	1255	738	450	1438
New Orleans, LA	756	915	1207	1487	1500	1697	1491	1439	1402	1258	952	804	1250
Caribou, ME	531	745	1192	1697	1771	1983	2015	1690	1343	937	450	391	1227
Boston, MA	454	745	1085	1328	1661	1823	1690	1582	1184	937	502	406	1110
E. Lansing, MI (low)	384	649	945	1284	1395	1638	1653	1432	1048	819	380	343	998
St. Cloud, MN	627	878	1461	1734	2070	2066	2118	1667	1343	1048	646	561	1352
Glasgow, MT	576	900	1146	1852	2362	2494	2435	2011	1395	900	642	450	1455
Lincoln, NB	686	930	1247	1576	1852	2052	2122	1775	1506	1114	771	613	1354
Las Vegas, NE	963	1292	1956	2111	2362	2771	2539	2332	2044	1483	1166	845	1822
Seabrook, NJ	686	908	1321	1668	1897	2007	1838	1771	1336	1052	771	535	1316
Albuquerque, NM	1133	1354	1834	2236	2494	2749	2502	2299	2018	1712	1284	1085	1802
New York, NY	450	705	956	1339	1572	1646	1620	1351	1166	897	546	395	1054
Hatteras, NC	941	1063	1550	2103	2229	2266	2229	2125	1587	1269	1015	756	1594
Cleveland, OH	373	675	1030	1550	2140	2214	2192	1934	1705	1041	487	406	1312
Stillwater, OK	923	1004	1520	1801	1838	2196	1889	1937	1565	1255	900	775	1467
Toronto, Ontario	351	605	1084	1317	1668	1926	1756	1627	1144	797	399	347	1078
Medford, OR	391	768	1232	2107	2790	2590	2804	2494	1553	1063	550	362	1575
State College, PA	506	642	1015	1428	1572	1845	1889	1683	1321	915	605	424	1154
Newport, RI	583	845	1196	1535	1786	1963	1860	1683	1358	1092	668	524	1238
Charleston, SC	923	1232	1664	2059	2288	2166	1989	1945	1509	1203	1139	786	1575
Nashville, TN	524	753	1089	1557	1838	1934	1867	1668	1439	1125	779	465	1253
El Paso, TX (high)	1328	1546	2125	2524	2716	2731	2531	2435	2403	1786	1428	1207	2037
Seattle, WA	229	328	1033	1823	1867	2286	2170	1753	1235	627	325	229	1160
Washington, D.C.	568	738	1225	1513	1716	1867	1808	1631	1373	1035	745	539	1234
Madison, WI	539	797	1166	1498	1727	1904	1993	1609	1321	959	557	443	1218

would total 63.1 million Btu's or 18,000 kwh. If about 20 percent of this energy is used to keep the digester at 95°F, then about 14,800 kwh would be available.

Designing the Physical Plant

Now that both energy needs and availability have been determined, a physical plant compatible with the community's needs must be designed.

a. Mechanical power needs (includes electricity):

Refrigeration	600 kwh/year
Irrigation	9,900 kwh/year
Grinding, mixing	231 kwh/year
Domestic water pumping	250 kwh/year
Lighting and small power	2,550 kwh/year
Miscellaneous small power	300 kwh/year
Total mechanical load	13,831 kwh/year

The mechanical load in New Mexico would be handled by wind power since water power is not available. One arrangement would be to have one 20-foot and one 15-foot wind generator close to the irrigation area providing a total of 10,906 kwh/year (see Table 30), or an auxiliary source could be provided to take the place of one of the wind generators. Another wind generator with a 15-foot propeller situated near the living area could supply the needs of the other five loads.

Table 30. Wind Power in Kw for Various Wind Speeds and Varying Diameter Blades, With 8,760 Hours in a Year

WIND	WIND GENERATOR BLADE DIAMETERS IN FEET					
SPEED	10	15	20	25	50	100
5	0.0247	0.0556	0.0987	0.154	0.617	2.47
10	0.197	0.455	0.790	1.23	4.94	19.8
15	0.666	1.50	2.67	4.16	16.8	66.6
20	1.58	3.55	6.32	9.87	39.5	158.0
25	3.09	6.94	12.3	19.3	78.1	308.5
30	5.33	12.0	21.3	33.3	133.4	533.1

b. Heat energy needs (solar energy source):

Space heating	108,000 kwh/year
Water heating	26,900 kwh/year
Cooking	1,200 kwh/year
Total heat energy load	136,100 kwh/year

For a flat-plate collector to supply the total energy load, its total area must be sufficient to collect the energy for the coldest month, January. The collector could be distributed over two or three dwellings, or if it is considered too large and expensive (collectors can cost from $1 to 2.50 and up per square foot), a small collector could be built and a conventional heating system installed to use wood, alcohol, or methane. This often proves to be cheaper and more efficient.

c. Chemical energy needs (methane, alcohol, wood):

Cooking	1,066 kwh/year
Cultivation (tractor)	5,600 kwh/year
Transportation	15,000 kwh/year
Total chemical loads	21,666 kwh/year

Depending upon the area of the Southwest considered, wood may or may not be too valuable to use as a fuel. The net available energy from methane in this location being only about 14,800 kwh, methane would supply only 70 percent of the total chemical energy needs. If an acre of sugar beets can be fermented and distilled into alcohol, it would provide up to an additional 6,600 kwh for various uses, though this figure is not the net yield of energy since some work will be expended in producing the extra crop. You also have to weigh the merits of using agricultural land for energy production instead of food.

ENERGY STORAGE

a. Mechanical energy storage (electrical in batteries, hydrogen, compressed air): The mechanical energy load which needs storage for use at random times of the day would be "lighting and small power" and "miscellaneous small power," totalling 2850 kw. About 8 kwh/day would be needed. If wind power supplies the mechanical energy, it would be wise to allow for three windless days storage, or 24 kwh. Therefore, the total capacity of a 120-volt battery system would have to be at least 200 ah (ampere-hours). ah × volt = wh (watt-hours).

b. Heat energy storage: One should store enough heat energy to take care of three cloudless days in the coldest month (January). For three days in New Mexico the total to be stored is 3×10^6 Btu. Different materials require different sized containers to provide this storage:

Water: 2,400 ft^3 necessary, with tank size 13.4 feet on a side
Rocks: 4,200 ft^3 necessary, with tank size 16 feet on a side
Salts: 280 ft^3 necessary, with tank size 6.5 feet on a side

c. Chemical storage: The storage of methane, alcohol, or wood is much easier than the other forms of energy. Wood, of course, can be piled into a shed. Methane can be compressed into cylinders or stored in inner tubes, floating water tanks, or porous materials like zeolites. Alcohol can be stored in tanks.

Load Sharing and Interconnection

Going back to the energy flow diagram (Figure 108), you will see the possible interconnections between energy sources, storage methods, and uses. This figure shows most of the possibilities of load sharing and interconnections. It is important to remember that no machine or device is 100 percent efficient in converting from one form of energy into another. Table 31 shows some very approximate efficiencies of various energy-producing devices.

Local energy resources in separate locales may be very different. The hypothetical community in the Southwest, for instance, would have an abundance of solar energy, which would then be used to compensate for a possible low wind speed, lack of water power, inability to raise many animals, and inability to grow wood on semi-arid land. In another part of the country, say, southern Wisconsin, solar insolation is low but this area has high average wind speeds. A large electric wind plant could be used for water heating, space heating, and charging electric cars and tractors.[16]

To summarize, the potential of a geographical area offers different energy forms which should be utilized and *combined* according to their availability. No region repeats the same factors as any other region. Always bear in mind that no single source will, in all likeli-

TABLE 31. Approximate Efficiencies of Various Energy-Related Devices

DEVICE	EFFICIENCY
Air compressor	50–60%
Air motor	85%
Direct burning of wood	29%
DC-AC inverter	85%
DC battery	75%
Electrolytic cell	60–90%
Electric generator	90%
Flat-plate collector	50–70%
Fuel cell	60%
Internal combustion engine	25%
Pelton water wheel	80–90%
Photosynthesis	0.2%
Solar heat engine	4–6%
Solar photovoltaic cell	10–14%
Water pump	65–85%
Water turbine	80–90%
Wind power plant	22–36%

hood, be adequate, and that combined sources should work in harmony with each other, making a whole that is greater than the sum of its parts.

What's Actually Being Done

Combined alternate energy systems are not merely hypothetical. Although very new in concept, a few combined systems have already been built, or are in the process of being built. For instance, near Albuquerque, New Mexico, Robert and Eileen Reines, with associates, have built a system that combines solar heating with electric power derived from the wind.[17] On the outskirts of the city of London, England, Bruce Haggart and Graham Caine have completed the initial construction of their "Street Farmhouse"—formerly called "Eco-House"—combining solar heat with agricultural and methane production and a wind-electricity system, showing that combination systems can be achieved even in a large city.[18]

Ms. A. N. Wilson of Martinsburg, West Virginia has achieved the combination of solar power, wind generation of electricity, and an aerobic composter to produce, if not methane, high-quality com-

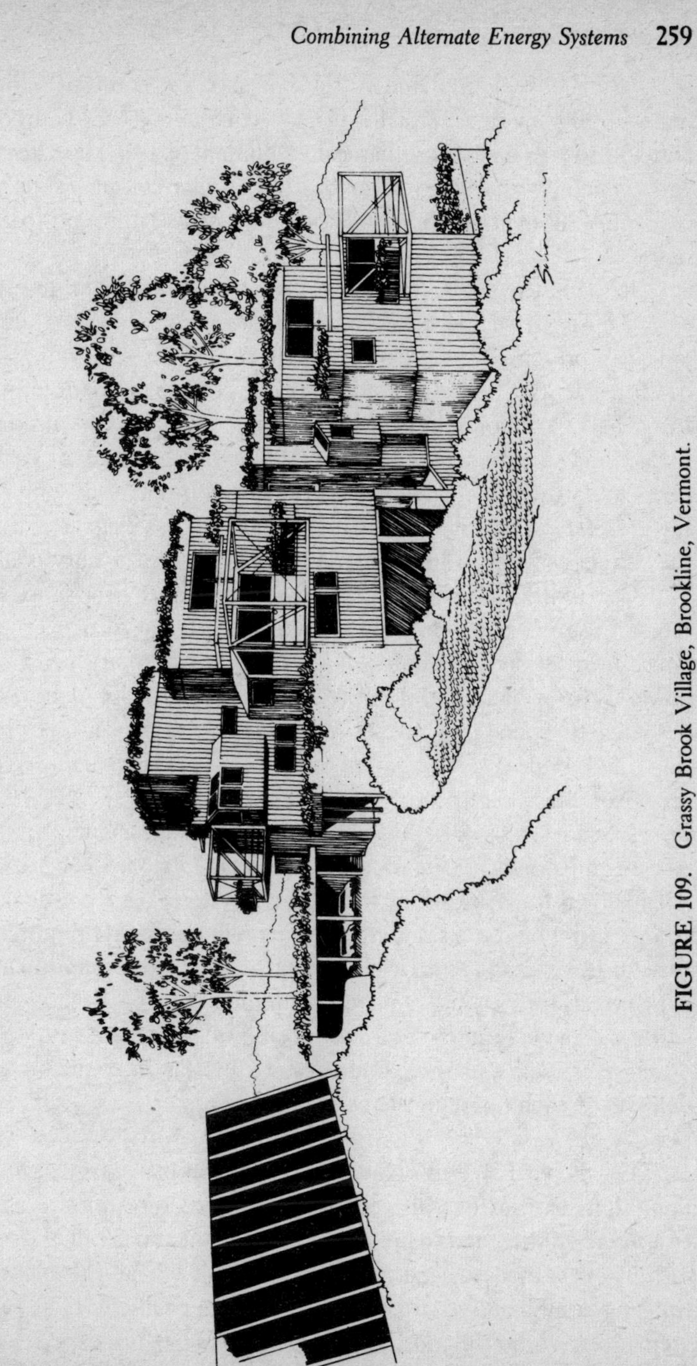

FIGURE 109. Grassy Brook Village, Brookline, Vermont.

post. Her house, completed in the fall of 1973, contains approximately 1400 square feet of living space (including four bedrooms) and 350 square feet of mechanical equipment space. Together, the wind system, with battery storage, and the solar complex supply 80 percent of all heating and power needed by this large, comfortable house.[19]

Perhaps the most ambitious project to date, is the one conceived by Richard Blazej of Newfane, Vermont.[20] "Grassy Brook Village," presently undergoing construction of Phase I, will consist of 20 residential houses—a small village, actually a condominium. But what a condominium! It is designed to make the minimum impact on the natural environment and provide life-support systems that derive their energy from natural and non-polluting sources to the greatest possible degree. A further goal of the village is to assure the resident-owners a high degree of financial and quality control.

Located on a 43-acre tract of woodland near Brookline, Vermont, Grassy Brook, when completed, will feature solar energy for house heating and hot water, generation of electricity from wind power, on-site handling of wastes in a system (still experimental) designed to operate in several recycling and pollution-free operations, and, hopefully, the production of methane for fuel. (See Figure 110.)

Life in Grassy Brook Village will be anything but primitive. Architect Robert F. Shannon and Project Engineer Fred Dubin (well-known for his work in solar energy systems), have designed a complex in which the standard of living will be, if anything, higher than in the average American suburb, because the design includes parks, vistas for viewing scenery, and minimal cost for utilities. Roof gardens provide beauty—and additional insulation. The complex, in other ways, will make maximum use of insulation, reducing heat demand of each house in the complex to approximately 7,500 Btu's per day.

Developer Blazej is organizing Grassy Brook Village as a condominium to provide the necessary legal structure for the use of common facilities, and to provide an opportunity for families to each own an ultra-modern, energy-saving dwelling. Of all the plans for utilizing combined systems for the benefit of people, Grassy Brook seems to us to be the most ambitious thus far. If the experimental waste treatment systems work well, Grassy Brook will be, within the

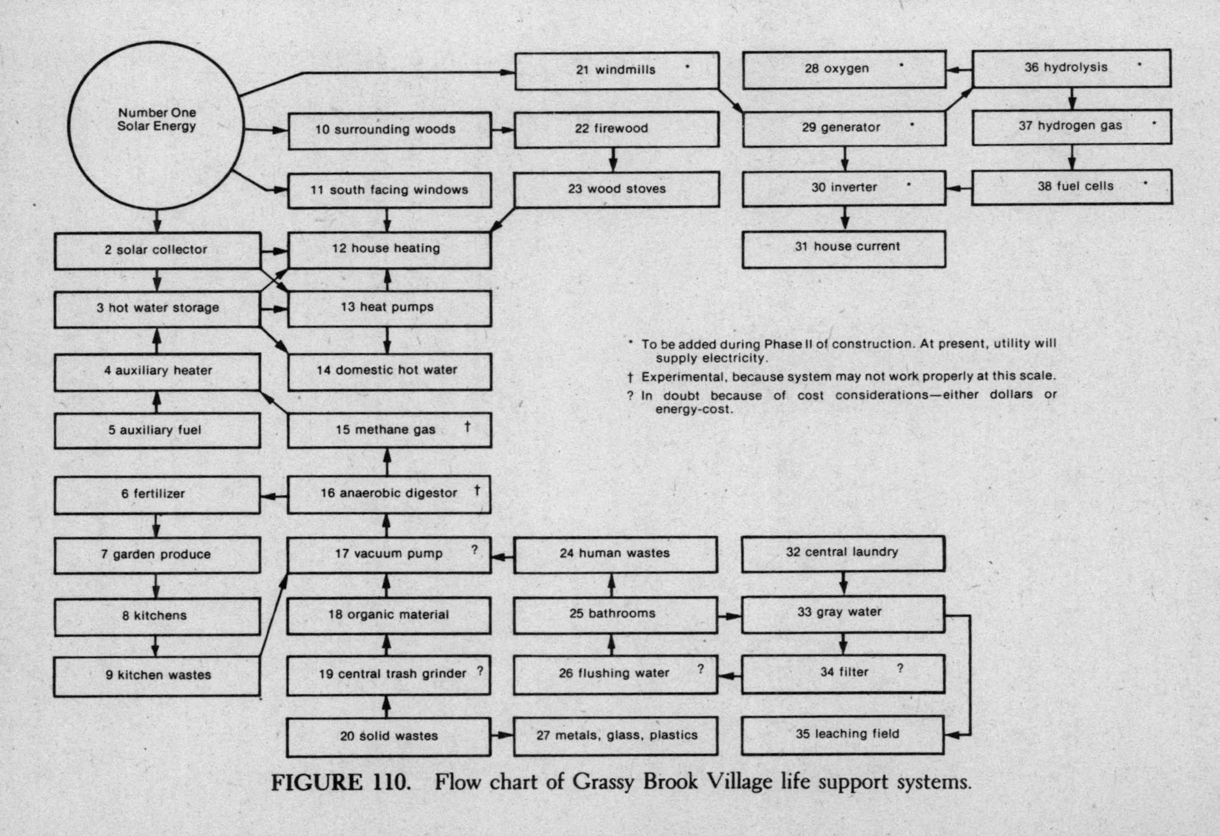

FIGURE 110. Flow chart of Grassy Brook Village life support systems.

next couple of years, the *first* village-scale application of mixed alternate energy systems—solar, wind, waste treatment, and methane fuel production.

APPENDIX II. CONSERVATION OF ENERGY IN EXISTING STRUCTURES

Eugene Eccli

Many of us are not in a situation where we can build a new house which might incorporate fuel-saving features, like heavy insulation and solar heating. We can take advantage of other possibilities, however. The costs are moderate and pay for themselves in a few years. In addition, even very old, uninsulated structures can save substantially on heat needs with a little "fixing up."

How a House Uses Heat

Let's look now at ways in which a structure (which could be a house, workshop, barn, or whatever) makes demands for heat. The greatest problem with some older structures is the influx of air through cracks and seams. We're all familiar with the uncomfortable drafts that come in around doors, windows, and other locations where the house isn't sealed well. The problem causes more than just discomfort, it also causes expense, because this infiltration of cold air must be heated.

SEALING

The first step in lowering the heating bill is to seal up the structure. Weatherstripping around doors and windows is a *must*. Caulking open cracks works best where there are no moving parts, such as around window and door frames. Caulking requires some effort but it is well worth it; where there were previously open drafts, even small ones, such sealing can save 10 percent of the heating bill.

Even with all doors and windows well sealed, quite a bit of cold air comes into the house by penetrating or migrating around the weatherstripping, or even directly through the walls. In the case of brick or concrete walls, the greater pressure of inside warm air, as well as the wind pressure against outside walls, can *force* air into your house. These inside-outside air exchanges can occur two or three times as often as is needed for fresh air comfort. Such exchanges add to the heating bill because all the fresh cold air must be heated. But there's also another problem involved. Since rapid changes in room air evaporate the moisture from our skin with a cooling effect, they also cause us to feel chilly—we must turn up the heat to achieve comfort.

The Biggest Offender—The Window

Several things can be done to relieve the situation. Luckily, the worst offender, the window, is the easiest problem to solve. Most windows are left closed during the winter. This is particularly true in older structures, since too much air gets in around windows and doors anyway. Adding the plastic sheet storm windows that are commercially available can do wonders for saving heat. The plastic stretches across the whole window and is held tight at the outer edges of the frame with either tacks or staples. The plastic is usually folded several times around a cardboard or paper strip; this bunching creates a near-airtight seal all around. (See Figure 111.)

The plastic, then, does several things. It slows down the in-and-out motion of air through the window area, getting rid of at least half the excess ventilation problem. It also serves as a dead-air space between window and plastic, slowing down the heat flow directly through the glass. I'd suggest putting the plastic sheets on the inside rather than the outside of the window. This is a bit unusual, but my experience has been that small animals (birds, field mice, cats) in search of warmth, rip through the plastic when it's on the outside, thus destroying the effectiveness of the barrier. Also, it's much easier to apply the plastic inside; no ladders are required to reach those second-story windows. For reasons I'll go into later, I also recommend that you leave 2 inches of window frame outside the perimeter of the plastic storm window.

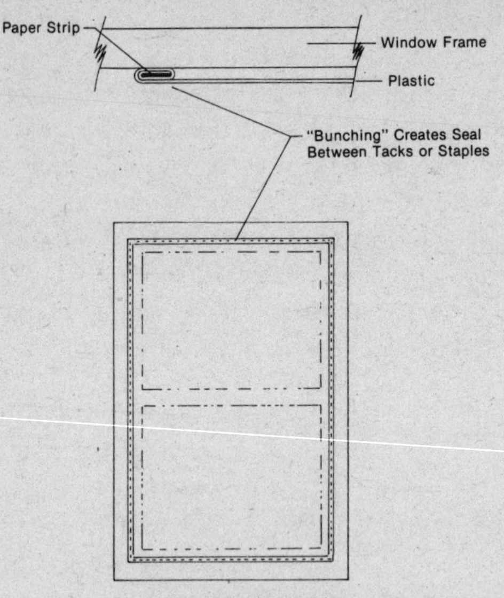

FIGURE 111. "Plastic storm windows" are most effective if the ends are folded around heavy paper strips before being tacked or stapled to window frames.

AIR COMING IN THROUGH WALLS

Stopping air motion through walls is a bit more difficult but if you plan to renovate an older structure, it may be worthwhile. A structure with bare brick or concrete walls can be painted with a sealer which will inhibit air motion as well as water penetration. Wood can also be sealed. A primer coat would be helpful, then apply an acrylic or tedlar-base paint to form a protective skin.

For the more adventurous who may be adding a wood or plasterboard finish to the inside of a brick or block wall, a plastic vapor barrier can be added before the finished surface is put up. Large sheets of plastic come in rolls which can be cut to any dimension and tacked around the perimeter of the walls in the same way that plastic windows are attached. This plastic will both inhibit the flow of air through walls and stop moisture (condensation from breath, cooking fluids, and other humidity) from penetrating from the inside

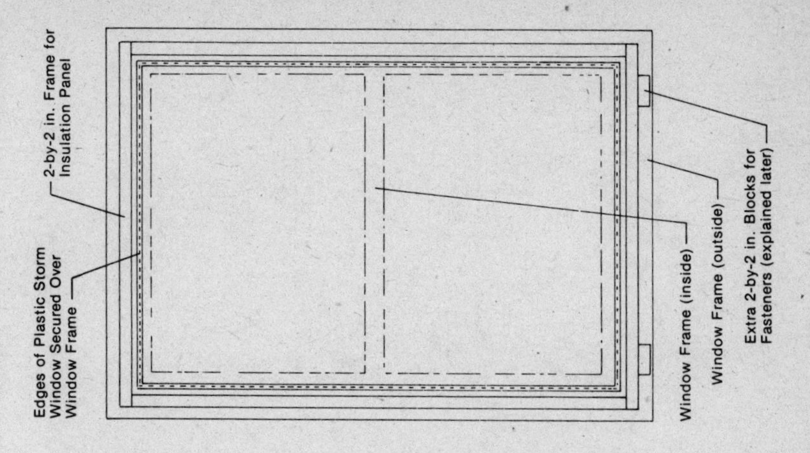

FIGURE 112. Fit your windows for insulation panels by nailing a 2-by-2-inch frame right over the original window frame.

of the house. This is important because a wet wall surface will lose heat to the outside much more quickly than a dry one.

Insulation

If we seal the structure moderately well, the next most important contribution to the heat bill is inadequate insulation, which allows heat to escape too rapidly through the walls, ceilings, floors, doors, and windows. Once again, the biggest offender is the window. More often than not, people use heavy curtains which they draw across the windows at night. Such action can be helpful, but a much more effective barrier to heat loss through window glass is a moveable insulation panel. This is fairly simple to make, shouldn't cost more than a good curtain, and is about five times as effective. Panels can be used on any window where movement (the panel raises to the ceiling during the day) doesn't interfere with furniture, shelves, or other fixed objects in the room. Here's how to make an insulation panel:

(1) We don't want the panel to interfere with the plastic over the inside of the window, so nail a 2-by-2-inch frame over the

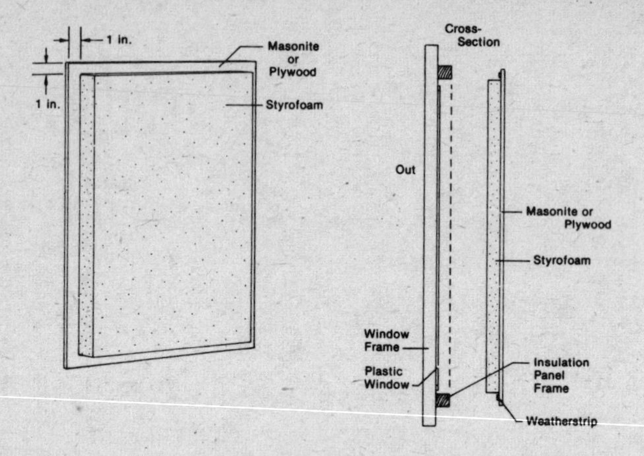

FIGURE 113. Cut the styrofoam to fit tightly inside the 2-by-2-inch frame, and cut the masonite or plywood larger so that it rests right on the frame.

window frame just outside the border of the plastic storm window. That's why I suggested that you not use up all the window-frame space when hanging your storm windows. (See Figure 112.)

(2) The panels themselves are made by gluing either ¼-inch plywood or masonite to a 1-inch slab of styrofoam. The styrofoam faces out-of-doors, the masonite or plywood, faces into the house. The styrofoam is cut to fit just inside the 2-by-2-inch frame, covering the entire window area. The plywood or masonite is cut about

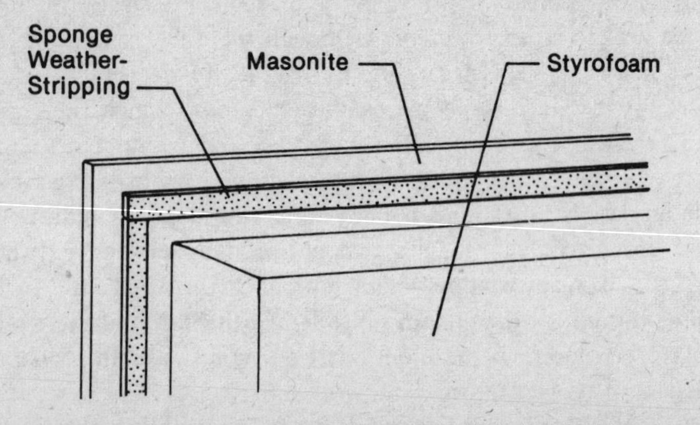

FIGURE 114. To create an airtight seal, place sponge weatherstripping on the masonite along the border of the 2-by-2-inch frame.

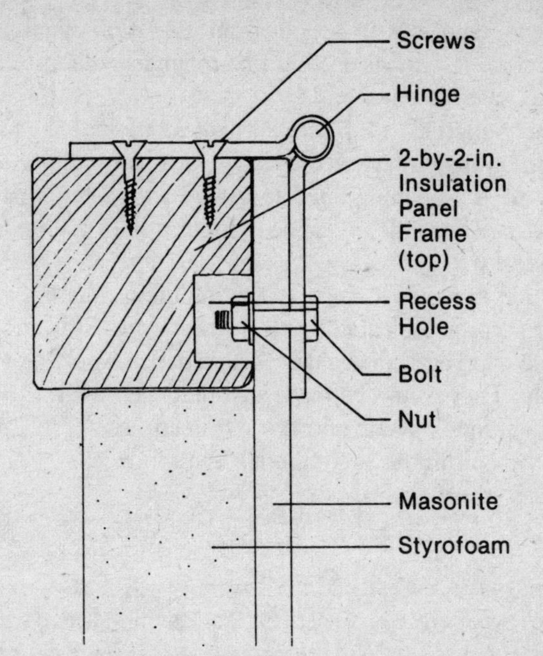

FIGURE 115. Drill recess holes in the frame and hinge the insulation panels to the masonite with bolts.

2 inches larger than the styrofoam on each side so that with the styrofoam glued on, the masonite can rest on the 2-by-2-inch frame. (See Figures 113a and b.)

(3) ¼-by-½-inch compressible (sponge) weatherstripping should be placed around the masonite lip so that when the panel is closed, a seal is made between the panel and the 2-by-2-inch frame. (See Figure 114.) This will stop virtually all air motion.

(4) The panel is then hinged to the masonite with short bolts; recess holes are drilled into the frame where the bolts come through the masonite. (See Figure 115.)

(5) Fasteners (see Figures 116a and b) can be placed on the bottom part of the 2-by-2-inch frame.

Of course, insulation panels are also most beneficial over regular storm windows and even over "thermo-pane" (two layers of glass with a small dead-air space between). Plastic storm windows are less efficient than these panels and much less efficient than thermo-pane. All systems of clear glass or plastic permit conductive heat loss out of the house, and *all* can benefit from insulation panels. Whether your structure is a mansion or a one-room schoolhouse, insulation panels will save you money.

Other variations are possible where rotation of the panel to its open position may interfere too much with lamps or furniture. For instance, interior-opening shutters can be constructed as easily as the insulation panels just described. (See Figures 117a and b.) Here's the general idea:

For larger windows such as the infamous "picture windows," construct accordion-pleated shutters (see Figures 118a and b) which would fold up in accordion shape. These panels would be made a bit differently. They would be made as a sandwich, with the masonite or plywood glued outside and the styrofoam inside. Hinges can be installed by bolting directly through the sandwich.

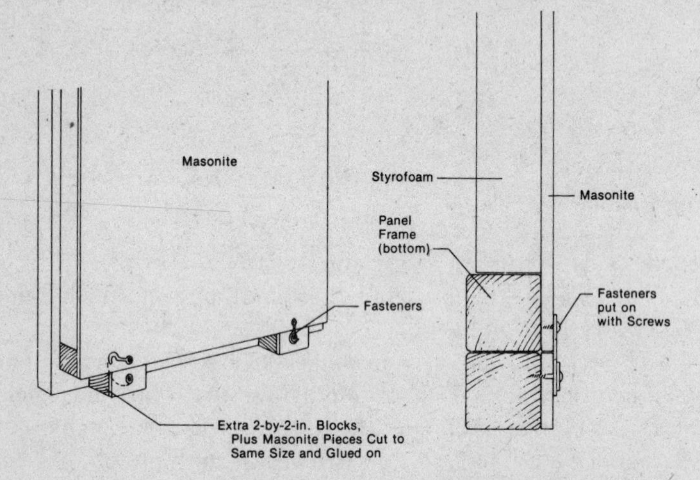

FIGURE 116. The panels can be secured at the bottom of the frame with fasteners. You'll need to glue onto the frame 2-inch blocks to make the panel surface and the frame surface even.

CEILINGS AND FLOORS

Next in line for improvement is an uninsulated ceiling. If there is a crawl space above the ceiling, 3 inches of fiberglass insulation will reduce heat needs in the house by 20 percent. Six inches of fiberglass would of course be better, if you can afford it.

Another important place to put insulation is under the floor. This can be done either from the cellar or by access to a crawl space. For either type of structure, it's important to seal the base of the house where there are cellar windows or an opening into the crawl space. Otherwise, a cold wind will suck out the heat from under the floor. Sealing windows and cracks here can be done as described earlier. An open crawl space under the house should be closed for the winter by placing a mound of dirt or sand bags all around the perimeter.

When considering fiberglass insulation for the ceiling or under the floor, be sure to buy the type with a reflective aluminum foil surface. When placing insulation, the shiny side of the fiberglass insulation should face the room's interior. This shiny surface reflects back the infrared heat waves which transfer heat out of the house. It can improve the effectiveness of insulation by about 10 percent if placed so it reflects heat back into the room. If possible, lay a sheet of plastic before placing insulation between joists. (See Figure 119.) The plastic will prevent moist house air from penetrating through to the insulation. Dry insulation is not only a much better heat-retainer, but also has a much longer life span.

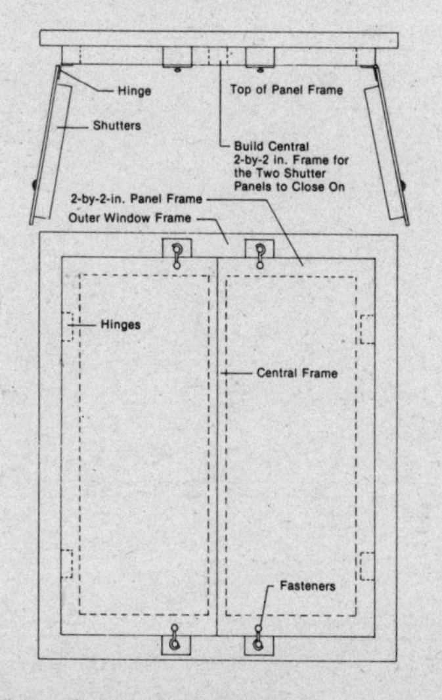

FIGURE 117. To make insulation shutters, construct two panels for each window and hinge them at the outer edges. Nail fasteners on the top and bottom as shown.

But what do you do if the crawl space above the ceiling or below the floor is too small to work in, or perhaps doesn't exist at all? For the ceiling, we can use styrofoam panels again. They're available in several sizes: 2-by-8-feet and 4-by-8-feet, for example. One inch of styrofoam is about as good as 2 of fiberglass, so having a 2-inch panel attached to the ceiling will serve about as well as 3 inches of fiberglass insulation above it.

The panels can be attached in several ways. For a flat surface, clean it and glue the panels in place. Another possibility is to use "roof buttons" (available at hardware stores) and nail the panels to ceiling studs. Where the ceiling is composed of rafters supporting the upper roof, the panels can be friction-fit between the rafters. A

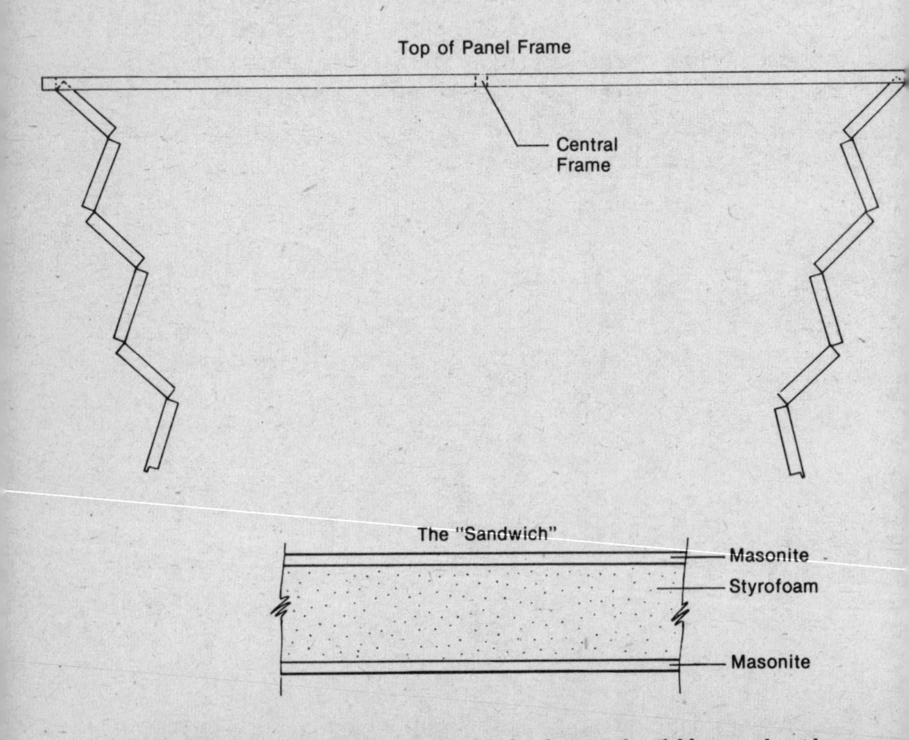

FIGURE 118. For larger windows make shutters that fold up at the sides by constructing many sandwich-like panels that are hinged to one another.

plastic vapor barrier would be unnecessary, by the way, as the styrofoam (itself a plastic substance) is impervious to water. A shiny aluminum surface to reflect heat back into the room can be made from cut-up sheets of 24-inch wide aluminum foil, stapled in place before the panels are attached. Even though such panels don't have the rustic wood look which some of us treasure in older houses, sealing and painting can do quite a bit to improve appearance.

Please note: precautions should be taken since styrofoam is flammable. The exposed surface should be well-coated with a fire-

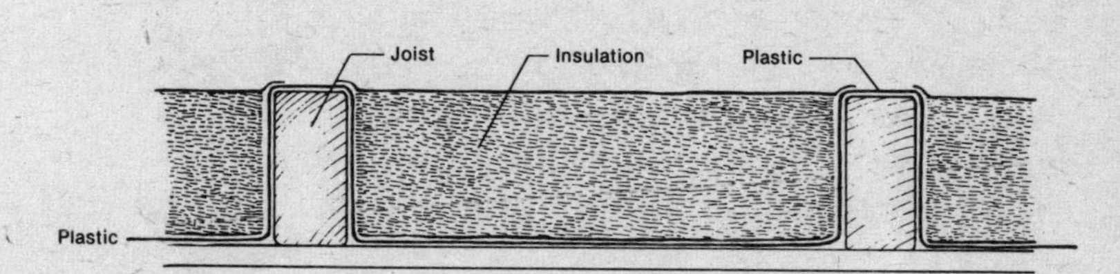

FIGURE 119. For best results when insulating floors and ceilings, place plastic sheets between joists and then staple insulation over the plastic with the reflective side facing into the room.

retardant paint. Pittsburgh Paint and other companies sell what is known as an "intumescent paint" which will do the job.

Where under-floor access is not possible, the easiest way to partially insulate the floor is by laying down heavy carpeting. But if you are going to renovate and put down new floors, the following may be possible: Place 2-by-2's over your old floor and nail them over the floor joists. Place 1½-inch styrofoam (painted, always with fireproof paint) in the spaces between the 2-by-2's (2-by-2's are actually sized 1½-by-1½ inches). Then lay your new floor over the 2-by-2-inch "joists."

Hot Air Circulation

Frequently people (even those with modern central heating) use several heating sources in various parts of the house. These include, among others, gas burners and fireplaces. The distribution of this heat through the house is important, especially if you wish to conserve energy and money by cutting 'way back on your central heating. What's more, good circulation eliminates the problem of some rooms being hot while others are cold. A single adequate heat source, with good circulation, can heat most areas of a house without the need for supplementary sources. This saves not only fuel, but initial investment in heating equipment.

The easiest ways to achieve circulation are the use of a fan and the floor-to-floor vent. Fans, of course, are included (as blowers) in all modern central heating plants, but you can achieve wonders from a small, inexpensive, low-speed electric fan (usually used for summer cooling), by using one to blow hot air into other rooms. The fan should blow air in a horizontal direction, and should be placed in a doorway, near the ceiling, since the ceiling is where heat builds up in a room that has a heat source. (See Figures 120a and b.) The fan should be coupled with open doors to other rooms where more heat is required.

For the upper floors, vents can be purchased at most hardware stores which allow air to rise from the lower to the upper stories. These vents come in different sizes, but the commonest is 8-by-10 inches. Cut a hole into the floor between the studs and insert the vent. Hot air, of course, rises from the lower to the upper floors through the vent, and cold heavy air from upstairs will drop down to be heated.

For larger houses, where some rooms need not be used, first close off all possible downstairs rooms, then the north rooms. Keep in mind that doing so may interrupt natural circulation patterns, which must then be reconstructed elsewhere through the house.

Since heat rises, the upper floor or floors will naturally be warmer if the air is given a chance to circulate upstairs. You can take advantage of the warmer temperature by arranging that many functions are performed upstairs. Another possibility might be to enclose stairways with a side wall and a door, so that with the stairway door closed during the day, little heat goes to the upper floor. With this arrangement, most of the heating during the day stays on the lower floor where people are working. At night, you can open the vents to allow most of the heat to go directly from the heat source to the upper floor. By closing some of the doors, the downstairs tempera-

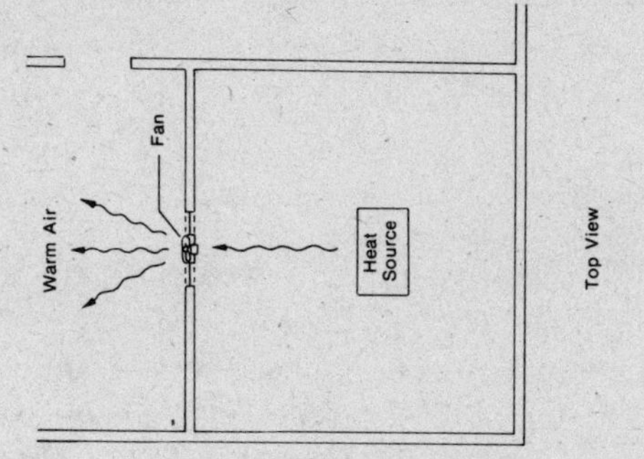

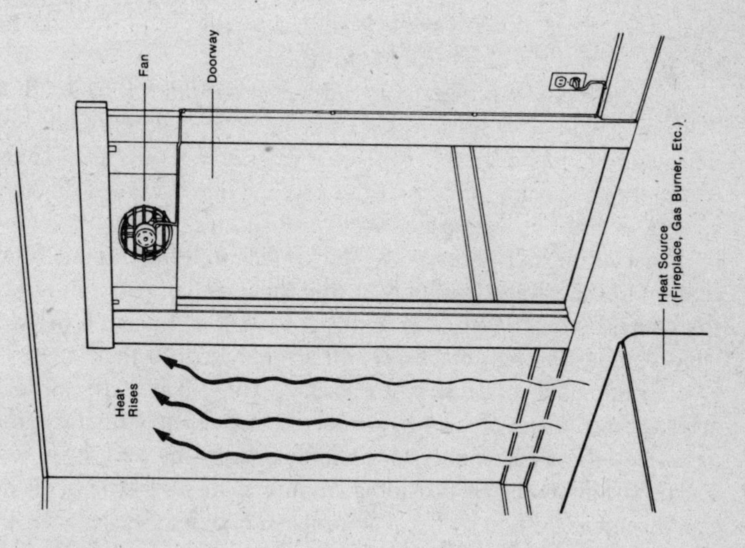

FIGURE 120. Small fans, placed in a doorway, can help in distributing the heat from gasburners, wood stoves, etc., to other parts of the house.

ture will fall at night. This method would keep the unused rooms at about 45°F on a cold night, and the occupied ones at 68°F, substantially saving on the heating bill.

North rooms get little sun and receive the largest chill from the wind, so if possible, cover the windows entirely with heavy curtains, or, better still, insulation panels; or close the room off altogether.

Some Heat-Saving Projects

AIR-LOCK ENTRANCE WAY

I mentioned earlier that a door as well as a window is a major area of cold air infiltration. Caulking and weatherstripping do only part of the job, because a door is opened several times during the day. Quite a bit of additional heat is needed in a house because each time we open a door, a lot of cold air enters. This is particularly bad when the door faces the prevailing winter winds on the north and west sides. First, if you have several doors, seal the north ones for the winter and use those on the south side.

A fairly simple project that can help even more is to build an "air-lock." This is just the old-fashioned foyer-type entranceway. The greatest air change that can occur when someone goes in or out is merely the volume of air in the foyer. Thus the most efficient air-lock is also the least expensive—a small one.

Insulation isn't necessary in the air-lock walls because its function isn't to stay warm, but only to stop the inrush of cold winds into the house. It should be well sealed to the outside, though, otherwise wind coming through cracks will force cold air into the house.

By using a foyer, the house door faces the still air in the air-lock and the usual losses of heat by air seepage would be virtually eliminated. So the air-lock saves heat in two ways: by very little loss around the door, and by providing a minimal change of air when we exit or enter the house. For additional use, a few shelves can be placed along one side of the foyer, as a place for coats and shoes. (See Figure 121.) The outer door, by the way, should open *out*, to ensure that in a fire no one could block the exit door.

The cost of an air-lock is small; scrap material can be used because of the small dimensions. Labor time shouldn't involve more than a few days. Ideas on how to frame a foyer can be found in

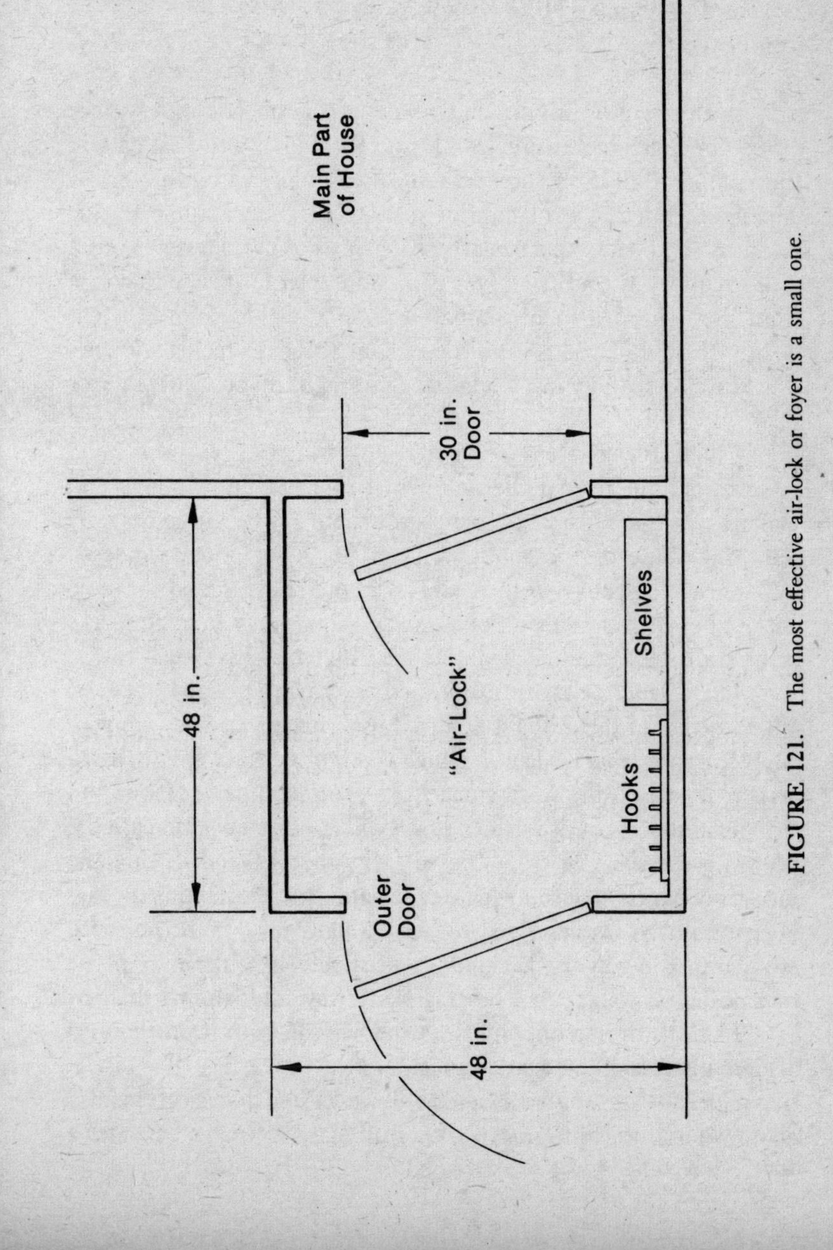

FIGURE 121. The most effective air-lock or foyer is a small one.

Fundamentals of Carpentry, Practical Construction, Vol. 2 (by Deerbahn and Sundberg, Chicago: American Technical Society).

USE OF FREE SOLAR HEAT THROUGH WINDOWS

Windows give us light and a view of the outside, but we pay a steep price for the luxury: heat loss in the winter and heat gain in the summer. If possible, close off some north windows for the winter. The light gained is small and the heat loss on the windy north side is really tremendous, particularly through windows. However, on a sunny winter day south windows gain more solar heat than they lose from the house. At night, though, the same windows lose all that heat to the out-of-doors—and then some! So, south-facing windows *must* be closed off by insulation panels at night if you want any free solar heat.

Taking fuller advantage of the sun's heat can be simple, and involves only increasing the number of south-facing windows and making moveable insulation panels for them. This can be done by removing some of the south wall and installing more windows. Where there are studs in the wall of a frame house, sections can be cut out between the studs. For a brick or concrete wall, some blocks or bricks have to be knocked out. (See Figures 122a and b.)

These windows can be made fixed or so that they can be opened and closed, and they should be fitted with inner plastic storm windows and moveable insulation panels, both of which were discussed earlier. If you do the work yourself, the cost will be very low, and the expansion of south window space will considerably increase the heat you receive free from the sun. By opening and closing the insulation panels on a sunny winter day and at the same time installing a small fan (as described earlier) to blow heat from the south room into the rest of the house, you should be able to turn the thermostat on your central heating plant 'way down during the day.

There is no reason why the windows can't run from floor to ceiling, increasing the window area (and the heat gain). The insulation panels—each of which would still weigh less than a refrigerator door—would be light enough to permit this. If weight becomes a

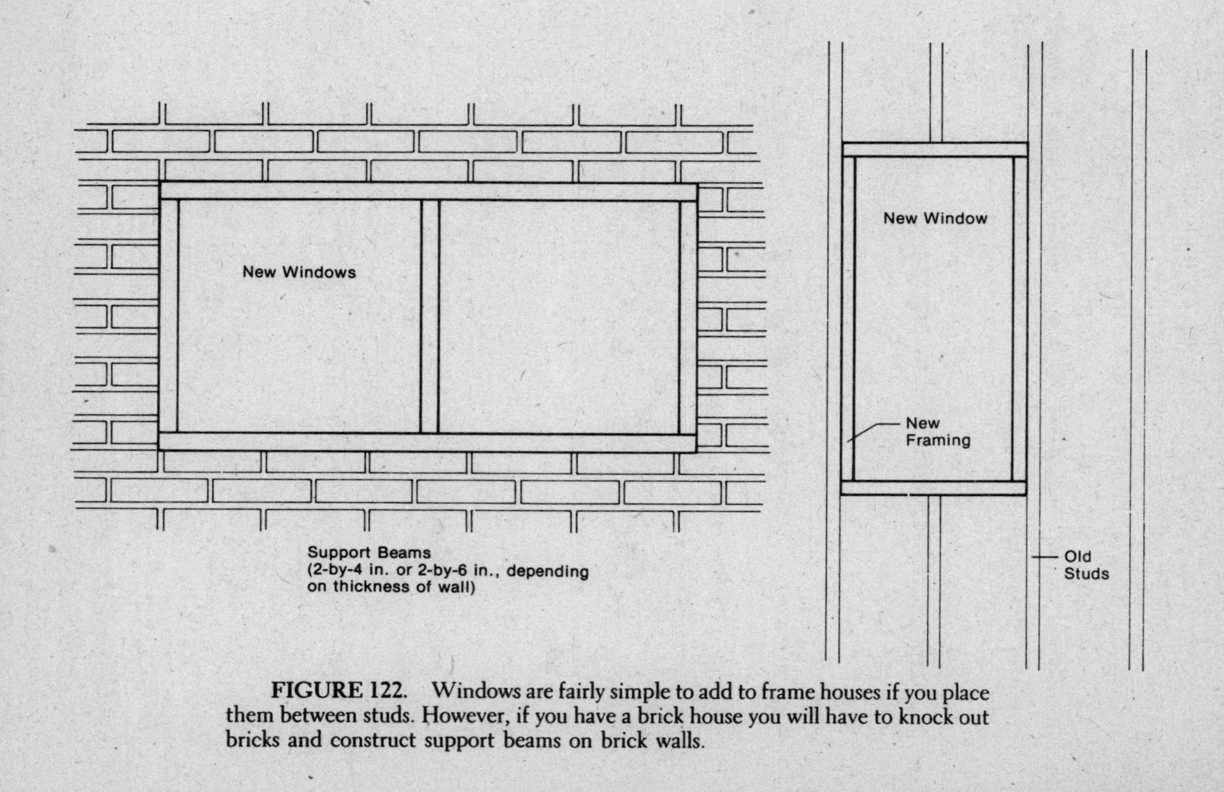

FIGURE 122. Windows are fairly simple to add to frame houses if you place them between studs. However, if you have a brick house you will have to knock out bricks and construct support beams on brick walls.

problem, the panels can be built as shutters that would swing out from either side (as described earlier).

Just how much heat can you expect from the sun? In most areas of the United States, the sun shines about 15 days during each winter month.* This is true except in very overcast areas such as the Pacific Northwest and parts of the Northeast such as much of New York State. For all other areas, each square foot of window space (when used in conjunction with moveable insulation panels) will

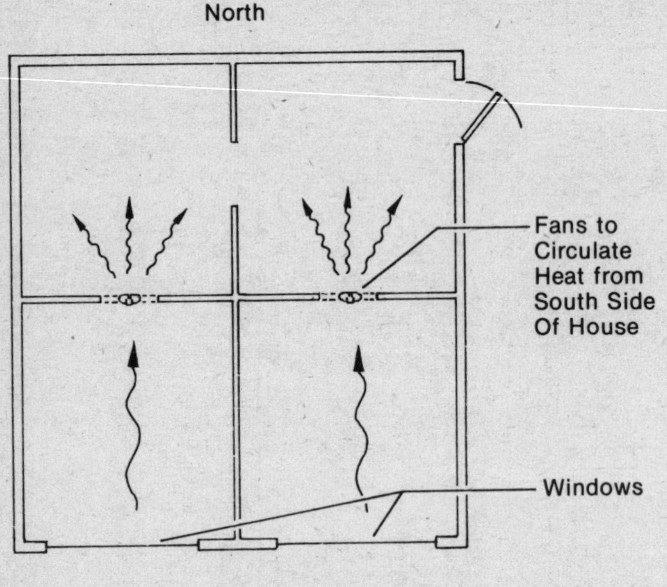

FIGURE 123. Conventional box-house structure.

heat one foot of floor space all winter. In a northerly climate, if you have 100 square feet of south-facing windows, you can heat a 100-square-foot room—for nothing! That's one less room to heat out of the average six-room house, hence the heating bill is reduced by 16 percent! For the sunny, warmer areas of the United States, the effect will be correspondingly greater.

---

*A sunny day is one that is less than 25 percent overcast.

House Construction

The idea of blowing heat from the south rooms to the rest of the house is really suited to the conventional box-structured house, where there are perhaps four rooms on each of two floors. Figure 123 shows this structure (one floor only), and the proper position of blower-fans to circulate heat from the south to the horth side of the house.

Some houses, however, are fortunate in having more of a south-facing orientation. That is, the shape is not square, but oblong with the long side facing south. (See Figure 124.) Here, air circulation is not necessary, as any heat which comes through the window will spread throughout the room on its own. In this case, what you may want to do is try to store some of the heat for use after the sun goes down. The easiest way to do this is to have a brick wall as the inner north wall of your house. The sun's radiant energy is absorbed by the reddish-brown bricks, heating them. Some of the heat seeps out to help heat the room during the day, but most is retained, and will serve as a heat source for the room at night. Such a system of full south-wall glazing, insulation panels, and a brick heat-storage wall can supply fully one-third of the heat in a house fortunate enough to have a long southern exposure!

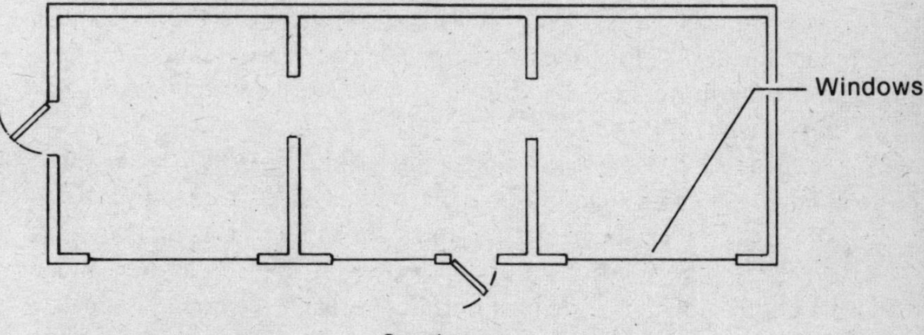

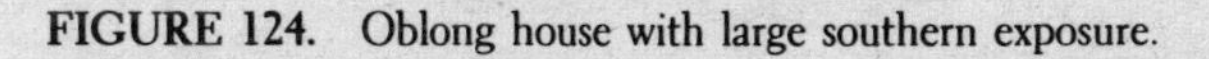
FIGURE 124. Oblong house with large southern exposure.

How can you add a brick wall in a house with standard walls? I'd start by reinforcing the floor with extra joists just below the place where the bricks will lie against the inner north wall. Next, attach

L-shaped "ties" on the old wall so that they stick out far enough to be cemented in with the bricks. This will tie the brick wall to the old wall to prevent its falling over. The brick wall is then built as usual, with one "tie" for each square foot of new brick wall. The bricks can be purchased locally, or some really valuable materials may be found where old houses are being torn down. These "recycled" bricks can be pieced together and the smaller bits can be used to make a very attractive mosaic. I won't describe the process, but *How to Work with Concrete and Masonry* by Darrell Huff (New York: Popular Science Publishing Co., Harper and Row) gives details for those unfamiliar with the skill of laying bricks. Won't the white mortar reflect some of the heat? True, but when mixing the mortar some dark dye can be added. It will give the wall a unique appearance and will also increase its heat-absorbing capacity slightly.

ADDITIONS TO THE HOUSE

Many people will be adding a new room to an already existing structure. Building that addition can offer several options for heating not just the new room but the house itself—if done properly.

The obvious first step is to place the addition onto the south side of the house. This gives the new room some protection from the wind, as well as allowing for maximum use of free solar heat. Any addition should, of course, make maximum use of insulation, with a minimum of 3 inches in the walls and 6 inches in the roof and under the floor. Then, the next step is to make maximum use of the south walls by building floor-to-ceiling windows and installing insulation panels.

This extra room can bring some real advantages. While the room is being used as a workshop, playroom, and/or storage area, the extra solar heat can be blown into the main house with a fan, helping to heat it. The addition can be closed off at night by a door and allowed to cool. The little heat escaping through the house wall will keep the addition from freezing, and the net effect will save fuel.

Not only doesn't the new addition have to be heated from the central plant, but it actually partially heats the house during sunny periods in winter. If use of the addition is required at any time when the sun is not shining, the fan can be reversed, and the normal house air will warm the addition. The moveable insulation panels and heavily-insulated construction of the addition will make this heat need very small.

Some new features can be added to this additional room. The air-lock idea can be incorporated into it, by building a small room between the main part of the house and the additional room. (See Figure 125.)

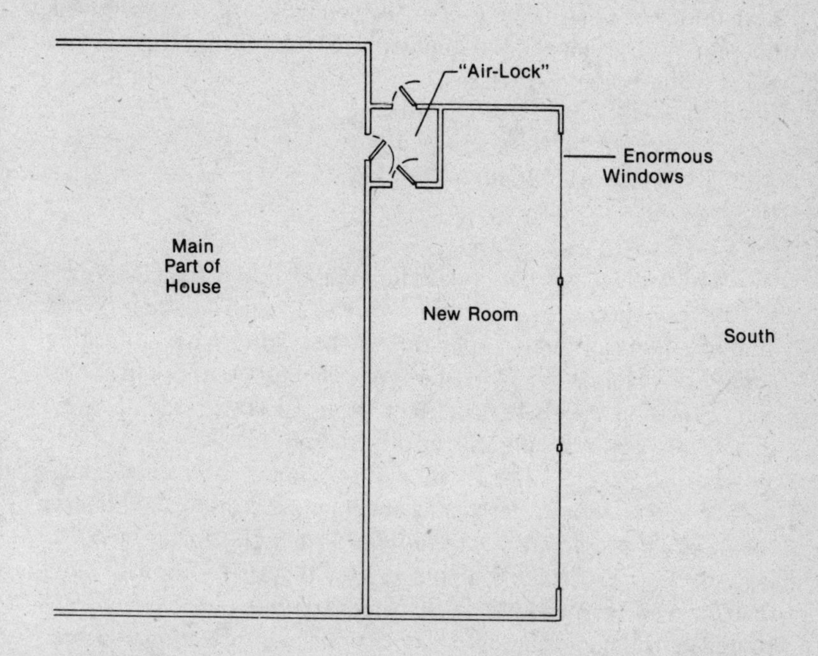

FIGURE 125. Large, south-facing windows with insulation panels, thick insulation; and an air-lock can make an addition of your house a real heat-saver.

Another use for the additional room is to make it, in part, a greenhouse. Usually a greenhouse requires more heat than any other room in a house or other structure. Even though it may soak up heat during a sunny winter day, the heat soon escapes through the glass after sunset, and large amounts of fuel are used in keeping the temperature up at night. Hence, the standard greenhouse addition is a heat-guzzler and gardeners pay a steep price for the hobby.

James Stamper of Roanoke, Virginia has come up with an idea that can change all this. If you have any amount of room in the main part of the house, simply roll the plants out of the greenhouse on moveable platforms at sunset. If there is a drop of a few steps into the greenhouse, install a ramp so that the "plant shelves" can be rolled in and out. This way, the greenhouse can be closed off at night, avoiding the heat demand while taking advantage of the sun's heat during the day. The excess heat problem (in relation to the house) can be avoided in the summer by venting the greenhouse and keeping the house door closed.

All This Will Cost Money—Will We Get It Back?

The average six-room, older house probably costs about $400 to $500 per winter to heat, and fuel prices are still going up. What kind of effect can the incorporation of these ideas have on heating costs? For a house without plastic windows, weatherstripping, moveable insulation panels and/or tight, heavy curtains, adding these should save about 25 percent of the heating bill.

For the basic addition of plastic windows, weathersealing, moveable insulation panels, and a small fan, the cost should be less than $125 *if you do the work yourself.* With a 25 percent saving in heat, these improvements should pay for themselves in little more than one winter! An additional 20 percent of heat can be saved every winter with the addition of 3 inches of ceiling insulation at a cost of about $50.

Building in some features may cost a little more, but again the savings come almost immediately. An air-lock, additional south windows and panels, and a brick wall or circulating fan (depending on the geometry of the house) shouldn't cost more than $200—again, if you do the work yourself.

In the already well-sealed square house described in the first part of this section, these built-in improvements will result in an additional 20 to 25 percent savings in heat. For the oblong house, there are more windows to purchase as the southern exposure is greater. This is an investment well made though, as the air-locks and solar heat system can save 40 percent on fuel bills. And these features certainly wouldn't hurt the resale value of the house.

I have not gone into too much detail regarding the construction of these improvements because there are reference books for the owner-builder which cover the necessary skills. None of the projects mentioned is beyond the capabilities of even the novice.

APPENDIX III. SOURCE LISTINGS

Wind Power

Aeromotor Div. of Brader Industries, 800 E. Dallas St., Broken Arrow, OK 74012. Manufacturer of water-pumping windmills.

Aircraft Components, Inc., North Shore Dr., Benton Harbor, MI 49022. Inexpensive, hand-held wind speed measuring equipment, including Dwyer and Taylor anemometers.

Automatic Power, Inc., Pennwalt Corp., 205 Hutcheson St., Houston, TX 77003. U.S. distributor for Aerowatt wind generators from France.

Bucknell Engineering Co., 10717 E. Rush St., S. El Monte, CA 91733. Wind-powered generating equipment.

Cramer Electronics, Inc., 85 Wells Ave., Newton, MA 02159. Large selection of electronic components.

Dempster Industries, Inc., 711 S. 6 St., Beatrice, NE 68310. Manufacturer of water-pumping windmills.

Domenico Sperandio, Via Cimarosa 13–21, 58022 Follonica (GR), Italy. Manufacturer of seven models, from 100 to 1,000 watts.

Dyna Technology, Inc., PO Box 3263, Sioux City, IA 51102. Twelve-volt Wincharger, which produces DC power.

Empire Electrical Co., 54 Mystic Ave., Medford, MA 02155. Source of special voltage and DC motors and generators.

Electro Sales Co., Inc., 100 Fells Way West, Somerville, MA 02145. Their catalog lists both new and government surplus rotary inverters, electric motors, and controls.

Environmental Energies, Inc., 11350 Schaefer St., Detroit, MI 48227. Windmill power plants.

W. W. Grainger, Inc., 5959 Howard St., Chicago, IL 60648. One-third horsepower electric motors.

Heath Co., Benton Harbor, MI 49022. Kit for a small DC inverter.

Heller-Aller Co., PO Box 25, Napoleon, OH 43545. Manufacturer of Baker windmills, pumps, water systems, tanks, and related supplies. (Baker wind units are for pumping water only, and are not adaptable for the generation of electricity.)

Henry Clews, Solar Wind Co., RFD #2, East Holden, ME 04429. Official northeastern states agent for Quirk's power plants and batteries (imported from

Australia); also distributes Elektro generators and blades (from Switzerland), and plants by Dunlite (from Australia) and Wincharger (a domestic make).

Laboratory for Maximum Potential Building Systems, School of Architecture and Planning, University of Texas, Austin, TX 78712. Consumer's guide to materials.

Lubing Machninenfabrik, Ludwig Benning, 2847 Barnstorf (Bez. Bremen), West Germany Postfach 171. Manufacturer of electricity-generating and water-pumping windmills.

Motorola Automotive Products, Inc., 9401 West Grand Ave., Franklyn Park, IL 60131. Transistor-regulated alternators.

Nova Manufacturing Co., 263 Hillside Ave., Nutley, NJ 07110. Their catalog lists over 1,000 electronic DC-to-AC inverters.

O'Brock Windmill Sales, Rte. 1, 12 St., North Benton, OH 44449. Windmills, towers, and repair parts from Baker, Dempster, Heller-Aller, and Aermotor (these units are for pumping water only); also, Davey hydraulic rams.

Palley Supply Co., 2263 E. Vernon Ave., Dept. M–70, Los Angeles, CA 90056. Source of surplus aircraft generators.

Popular Mechanics, Dept. C, Box 1014, Radio City, NY 10019. Plans and instructions for the construction of wind generators, air-brake governors, and slip-ring commutators.

Sensenich Corp., PO Box 1168, Lancaster, PA 17604. This manufacturer of wooden propellers is interested in working with wind power plant innovators and generator manufacturers.

Gerry Smith, University of Cambridge, 1 Scroope Terrace, Cambridge CB 2 1 PX, England. Has compiled an international list of windmill manufacturers, past and present.

Surplus Center, 1000 W. O St., PO Box 82209, Lincoln, NB 68501. Parts for wind power projects.

Water Power

James Leffel & Co., Springfield, OH 45501. A leading manufacturer of small (500- to 10,000-watt) hydroelectric power plants. They stock the Hoppes Hydro-Electric Unit, which can be used to generate AC or DC current, or can be used for the direct drive of machinery such as pumps, grist mills, etc. Bulletin H–49 describes the unit, and gives instructions for measuring the quantity of water in a stream.

O'Brock Windmill Sales, Rte. 1, 12 St., N. Benton, OH 44449. Davey hydraulic rams.

Rife Hydraulic Engine Mfg. Co., Box 367, Millburn, NJ 07041. Hydraulic rams.

Surplus Center, 1000 W. O St., PO Box 82209, Lincoln, NB 68501. Parts for water power projects.

Wood Power

Ashley Automatic Heater Co., PO Box 730, Sheffield, AL 35660. Their catalog lists many wood-burning designs.

Black's, 58 Maine St., Brunswick, ME 04011. Dealer for the Shenandoah wood stove (for auxiliary heat and for cabins) and the Maine Barrel Stove (for auxiliary heat).

Emil Dahlquist, 31 Morgan Park, Clinton, CT 06413. Manufactures a fireplace rotisserie and a tin fireplace oven. These designs are used with a special grate that is raised at the front, permitting heat from the coals to radiate out into the room.

Extension Forester, Fernow Hall, Cornell University, Ithaca, NY 14850. This office has a list of manufacturers of wood-burning units.

Loeffler Heating Units Mfg. Co., RD 1, Box 503-0, Branchville, NJ 07826. A movable hot-air heating device for fireplaces that uses an electric blower to drive air through a steel pipe grate and into the room.

The Majestic Co., Huntington, IN 46750. One of America's leading fireplace manufacturers. Some models incorporate ducts that circulate warm air.

Monumental Millwork, Inc., 3500 Parkdale Ave., Baltimore, MD 21211. Distributor of Heatilator fireplace units for Maryland, Pennsylvania, Delaware, Virginia, and West Virginia. Some Heatilator models feature warm air ducts that can be used to heat several rooms. Available by mail order.

Mother's Truck Store Catalog, Box 75, Unionville, OH 44088. A catalog listing many different kinds of new wood stoves.

Patented Manufacturing Co., Lincoln Rd., S. Lincoln, MA 01773. Manufacturer of a simple stovepipe heat exchanger that uses metal fins to radiate heat.

Portland Franklin Stove Foundry, Inc., Box 1156, Portland, ME 04104. Manufacturer of Queen Atlantic wood ranges, Home Atlantic wood parlor stoves, and Franklin stoves. Available by mail order.

Preston Distributing Co., Whidden St., Lowell, MA 01852. Dealer of coal and wood burning furnaces and stoves, and New England distributor of fireplace coal.

Riteway Manufacturing Co., Div. of Sarco Corp., PO Box 6, Harrisonburg, VA 22801. Manufacturer of thermostat-controlled wood and coal furnaces and boilers. Available by mail order.

Shenandoah Manufacturing Co., PO Box 839, Harrisonburg, VA 22801. Manufacturer of wood burning brooders which, according to the company, can function as furnaces for small dwellings. Available by mail order.

Superior Fireplace Co., 4325 Artesia Ave., Fullerton, CA 92633. Manufacturer of Heatform warm-air-circulating fireplaces.

Vega Industries, Inc., Mt. Pleasant, IA 52641. Manufacturer of Heatilator fireplaces.

Waverly Heating Co., Beverly, MA 01915. The largest wholesaler of wood stoves in New England.

Weir Stove Co., Taunton, MA 02780. The last original stove manufacturer in New England.

Methane Power

AB Clivus, Tonstigen 6, S–135 00 Tyreso, Sweden. Manufacturers of the Clivus, a household organic waste digester.

Conrad Budny, 537 Mantua Ave., Woodbury, NJ 08096. Budny converts cars to run on either methane or gasoline.

Earth Move, Box 252, Winchester, MA 01890. A methane conversion kit for automobiles that reduces the pressure of compressed methane gas to a level usable in a carburetor. No special tools or experience are required to make the conversion.

Lakeside Equipment Corp., 1022 E. Devon Ave., Bartlett, IL 60103. Manufacturer of the Spiragester which, according to *Solar Energy Digest*, can be used as a methane gas producer.

Solar Power

Arkla Industries, Inc., 950 E. Virginia St., Evansville, IN 47701. Manufacturer of absorption air conditioning equipment. Some models are adaptable to produce cooling, when powered by water heated in flat plate collectors.

Beutel Solar Heater Co., 1527 N. Miami Ave., Miami, FL 33132. Manufacturer of solar water heaters.

Edward's Engineering Corp., 101 Alexander Ave., Pompton Plains, NJ 07444. Manufacturer of a complete system for heating, air conditioning, and water heating.

Enthone, Inc., Box 1900, New Haven, CT 06508. Aqueous bath solution for blackening the steel surfaces of flat plate collectors.

Freeman A. Ford, President, FAFCO, 2860 Spring St., Redwood City, CA 94063. Solar pool heaters.

Fun and Frolic, Inc., Dept. S, PO Box 277, Madison Heights, MI 48071. Maker of the Solorator, an inexpensive flat plate solar swimming pool heater.

Helio Associates, Inc., 8230 E. Broadway, Tucson, AZ 85710. Plans for a solar energy heater.

Don Johnson, 2523 16 Ave. S., Minneapolis, MN 55404. Sells 49-inch parabolic reflectors.

Kalwall Corp., PO Box 237, Manchester, NH 03105. Manufacturer of polyester reinforced fiberglass sheet for use as solar collector covers. The material is lightweight and has excellent light transmittance. The company has developed heat-siphoning, hot-air producing panels, and is interested in manufacturing these units on an experimental basis.

Ram Products Co., 1111 N. Centerville Rd., Sturgis, MI 49091. "Plexi-view" acrylic mirrors that are lighter than glass and have excellent optical qualities.

Sky Therm Processes and Engineering, 945 Wilshire Blvd., Los Angeles, CA 90017. Has developed a modular roof system, in which panels of movable insulation allow water beds in the ceiling to absorb heat from either the outside or the rooms below.

Solar Energy Co., 810 18 St., NW, Washington D.C. 20006. Six- and 12-volt solar cell battery chargers.

Solar Energy Digest, PO Box 17776, San Diego, CA 92117. U.S. distributor for Solapak solar water heaters, made in Australia.

Suhay Enterprises, 600 Ivy St., Glendale, CA 91204. An inexpensive solar water heater that will contain, heat, and insulate over four gallons of water.

Sunworks, Inc., 669 Boston Post Rd., Guilford, CT 06437. Flat plate solar heat collectors, both surface-mounted and flush-mounted. The company also has technical details, efficiency graphs, detailed drawings, and other information.

Thomason Solar Homes, Inc., 6802 Walker Mill Rd., SE, Washington D.C. 20027. Maker of the Thomason Solaris System for home heating and cooling. The company conducts seminars, and licenses companies and individuals wishing to practice the Thomason inventions. Their plans for solar houses are available from this address: Edmund Scientific Co., 605 Edscorp Building, Barrington, NJ 08007.

Transparent Products Corp., 1727 W. Pico Blvd., Los Angeles, CA 90015. Aluminized mylar, a plastic sheet that is coated on one side with a reflective aluminum finish, suitable for use in making concentration collectors.

Tranter Manufacturing, Inc., 735 E. Hazel St., Lansing, MI 48909. Maker of Econocoil, a stainless steel flat plate collector.

APPENDIX IV. CONVERSION TABLES

Units of Length

1 mile	= 1760 yards	= 5280 feet
1 kilometer	= 1000 meters	= 0.6214 mile
1 mile	= 1.607 kilometers	
1 foot	= 0.3048 meter	
1 meter	= 3.2808 feet	= 39.37 inches
1 inch	= 2.54 centimeters	
1 centimeter	= 0.3937 inch	

Units of Area

1 square mile	= 640 acres	= 2.5899 square kilometers
1 square kilometer	= 1,000,000 sq. meters	= 0.3861 square mile
1 acre	= 43,560 square feet	
1 square foot	= 144 square inches	= 0.0929 square meter
1 square inch	= 6.452 square centimeters	
1 square meter	= 10.764 square feet	
1 square centimeter	= 0.155 square inch	

Units of Volume

1 cubic foot	= 1728 cubic inches	= 7.48 U.S. gallons
1 british imperial gallon	= 1.2 U.S. gallons	
1 cubic meter	= 35.314 cubic feet	= 264.2 U.S. gallons
1 liter	= 1000 cubic centimeters	= 0.2642 U.S. gallons

Units of Weight

1 metric ton	= 1000 kilograms	= 2204.6 pounds
1 kilogram	= 1000 grams	= 2.2046 pounds
1 short ton	= 2000 pounds	

Units of Pressure

1 pound per square inch (psi)	= 144 pound per square foot
1 pound per square inch	= 27.7 inches of water*
1 pound per square inch	= 2.31 feet of water*
1 pound per square inch	= 2.042 inches of mercury*
1 atmosphere	= 14.7 pounds per square inch (psi)
1 atmosphere	= 33.95 feet of water*
1 foot of water = 0.433 psi	= 62.355 pounds per square foot
1 kilogram per square centimeter	= 14.223 pounds per square inch
1 pound per square inch	= 0.0703 kilogram per square centimeter

*at 62°F (16.6°C)

Units of Power

1 horsepower (English)	= 746 watts = 0.746 kilowatts
1 horsepower (English)	= 550 foot pounds per second
1 horsepower (English)	= 33,000 foot pounds per minute
1 kilowatt	= .948 British thermal units (Btu)/-second

1 kilowatt (kw) = 1000 watt	= 1.34 horsepower (hp) English
1 horsepower (English)	= 1.0139 metric horsepower (cheval-vapeur)
1 metric horsepower	= 75 meter × kilogram/second
1 metric horsepower	= 0.736 kilowatt = 736 watt

NOTES

Wind Power

1. *Proceedings of the United Nations Conference on New Sources of Energy*, Volume 7, *Wind Power.* (New York: U.N. Press, 1964) 408 pp.

2. Palmer C. Putnam, *Power From the Wind.* (New York: VanNostrand & Co., 1948).

3. National Weather Records Center, U.S. Weather Bureau, Federal Building, Asheville, NC 28801. Statistical weather data for all of the U.S. for the past 50 years.

4. Electro Sales Co., Inc., 100 Fellsway West, Somerville, MA 02145. Their 200-page catalog lists new and government surplus rotary inverters, electric motors, and controls.

5. Aircraft Components, Inc., North Shore Dr., Benton Harbor, MI 49022, has inexpensive hand-held models. It also carries wind speed measuring equipment including Dwyer and Taylor anemometers.

6. Robert Reines, "Wind Energy," *Life Support Technics Conference Proceedings*, October 31, 1972, Ghost Ranch, Abiquiu, NM, p. 200.

7. S. Kidd and D. Garr, "Electric Power from Windmills?" *Popular Science.* (November 1972) p. 72.

8. E. W. Golding, *The Generation of Electricity by Wind Power.* (New York: Philosophical Library, 1956), p. 235.

9. Reines, "Wind Energy," p. 20.

10. Golding, *Generation of Electricity*, p. 210.

11. *Ibid.*, p. 58.

12. Information taken from Quirk's tower installation pamphlet.

13. J. Sencenbaugh, in personal letter to the author.

Wood Power

1. Ken Kern, *The Owner-Built Home.* Available from the author at Sierra Route, Oakhurst, CA 93644.

2. This efficient wood furnace is manufactured by Sarco Corporation, Box 6, Harrisonburg, VA 22801. The furnace, however, is now quite expensive and costs several thousand dollars. Even their medium-sized wood heater sells for a few hundred dollars.

3. Conversion doors and legs for these barrel stoves can be obtained from Washington Stove Works, Everett, WA 98206.

4. The Nippa wood-burning sauna heater is available from Bruce Mfg. Co., Bruce Crossing, MI 49912.

Methane Power

1. L. Anderson, *Energy Potential from Organic Wastes: A Review of the Quantities and Sources.* Bureau of Mines Information Circular 8549, U.S. Dept. of Interior, 1972.

2. M. Perelman, "Efficiency in Agriculture: The Economics of Energy," in *Radical Agriculture,* edited by Richard Merrill (New York: Harper & Row), forthcoming.

3. C. Acharya, *Preparation of Fuel Gas and Manure by Anaerobic Fermentation of Organic Materials.* ICAR Research Series No. 15 (New Delhi, India: Indian Council of Agricultural Research, 1958), 58 pp.

4. C. Acharya, *Organic Manures.* Research Revision Series, Bulletin No. 2 (New Delhi, India: Indian Council of Agricultural Research, 1952).

5. Ram Bux Singh, *Bio-Gas Plant* (Ajitiman, Etawah [U.P.], India: Gobar Gas Research Station, 1971).

6. Ram Bux Singh, *Some Experiments with Bio-Gas* (Ajitimal, Etawah [U.P.], India: Gobar Gas Research Station, 1971).

7. K. Imhoff and G. Fair, *Sewage Treatment* (New York: John Wiley & Sons, Inc., 1956).

8. K. Imhoff, W. Muller, and D. Thistlewayle, *Disposal of Sewage and Other Water-Borne Wastes* (Ann Arbor: Ann Arbor Science Publishers, Inc., 1971).

9. M.S. Anderson, *Sewage Sludge for Soil Improvement.* USDA Circular No. 972, 1955.

10. James P. Law, *Agricultural Utilization of Sewage Effluent and Sludge.* Annotated bibliography, Federal Water Pollution Control Administration, U.S. Dept. of Interior, 1968.

11. S. Greeley and C.R. Velzy, "Operation of Sludge Gas Engines," *Sewage Works Journal 8* (1936):57–62.

12. J. Jeris and P. McCarty, "The Biochemistry of Methane Fermentation Using C^{14} Traces," *Journal of the Water Pollution Control Federation* 37(1965):178–192.

13. A.M. Boswell, "Microbiology and Theory of Anaerobic Digestion," *Sewage Works Journal* 19(1947):28.

14. H. Barker, *Bacterial Fermentation* (New York: John Wiley & Sons, 1956).

15. American Public Health Association, *Standard Methods for the Examination of Water and Wastewater.* 11 ed. (New York, 1960).

16. Gordon L. Dugan, et. al., *Photosynthetic Reclamation of Agricultural Solid & Liquid Wastes.* SERL Report No. 70–1 (University of California at Berkeley: Sanitary Engineering Research Laboratory, 1970).

17. Eliseos P. Taiganides and T. Hazen, "Properties of Farm Animal Excreta," *Transamerican Society of Agricultural Engineering* 9(1966):374–376.

18. C.F. Schnabel, "A Voice of Eco-Agriculture," *Grass Acres.* March, 1973. Box 1456, Kansas City, MO. (A monthly newspaper of high quality.)

19. E.P. Taiganides et. al., "Anaerobic Digestion of Hog Wastes," *Journal of Agricultural Engineering Research* 8(4).

20. Schnabel, "A Voice of Eco-Agriculture."

21. Perelman, "Efficiency in Agriculture."

22. S. Klein, "Anaerobic Digestiön of Solid Wastes," *Compost Science.* February 1972.

23. R. Laura and M. Indai, "Increased Production of Bio-Gas from Cow Dung by Adding Other Agricultural Waste Materials," *Journal of the Science of Food and Agriculture* 22(1971):164–167.

24. Klein, "Anaerboic Digestion."

25. J. Patel, "Digestion of Waste Organic Matter and Organic Fertilizer and a New Economic Apparatus for Small Scale Digestion," *Poona Agricultural College Magazine* (India) 42(1951):150–159.

26. John Fry, "Manure Smell Furnishes Farmstead's Power Needs," *National Hog Farmer* 6(1961):3.

27. John Fry, "Power and Electric Light from Pig Manure," *Farm and Country* (London) April 1960.

28. E.P. Taiganides, "Characteristics and Treatment of Wastes from a Confinement Hog Production Unit," Ph.D. dissertation, Iowa State University Science and Technology, Agricultural Engineering, 1963. Available from University Microfilm, Inc., Ann Arbor, Michigan, No. 63–5200, 177 pp.

29. Fry, "Manure Smell."

30. Fry, "Power and Electric Light."

31. Acharya, "Preparation of Fuel Gas."

32. Singh, *Bio-Gas Plant.*

33. Singh, *Some Experiments.*

34. Dugan, "Photosynthetic Reclamation."

35. "Plowboy Interview with Ram Bux Singh," *Mother Earth News* 18(1972):7–11.

36. F.H. King, *Farmers of Forty Centuries: Permanent Agriculture in China, Korea, and Japan* (Emmaus, Pa: Rodale Press, Inc., 1973).

37. J. Scott, *Health and Agriculture in China* (London: Faber & Faber, 1952).

38. H. Gotaas, *Composting: Sanitary Disposal and Reclamation of Organic Wastes* (Geneva, Switzerland: World Health Organization, 1956).

39. L. Hills, "The Clivus Toilet—Sanitation Without Pollution," *Compost Science.* May/June 1972.

40. Imhoff, Muller, and Thistlethwayle, *Disposal of Sewage.*

41. F. Schmidt and W. Eggergluess, "Gas from Agricultural Waste," *Gas Journal* 279(1954):2861.

42. G. Rosenberg, "Methane Production from Farm Wastes as a Source of Tractor Fuel," *Journal of the Ministry of Agriculture* (England) 58(1952):487–94.

43. K. Imhoff and C. Keefer, "Sludge Gas a Fuel for Motor Vehicles," *Water Sewage Works* 99(1952):284.

44. John Fry, "Farmer Turns Pig Manure into Horse-Power," *Farmer's Weekly* February 22, 1961 (Bloemfoutem, South Africa), p 16.

45. Imhoff and Fair, *Sewage Treatment.*

46. Ministry of Agriculture, *Fisheries, and Food, Nitrogen and Soil Organic Matter* (London: National Agriculture Advisory Service, Her Majesty's Stationery Office, 1964.)

47. E. Coker, "The Value of Liquid Digested Sludge," *Journal of Agricultural Science* 67(1966):91–97.

48. Imhoff and Fair, *Sewage Treatment.*

49. Taiganides, *Characteristics and Treatment of Wastes.*

50. A. Wolff and A. Wasserman, "Nitrates, Nitrites, and Nitrosamines," *Science* 177(1972):15–18.

51. H. Walters, "Nitrate in Soil, Plants, and Animals," *Journal of the Soil Association* 16(1970):1–22.

52. James P. Law, "Nutrient Removal from Enriched Waste Effluent by the Hydroponic Culture of Cool Season Grasses," Federal Waste Quality Administration, Department of Interior, 1969.

53. H. Iby, "Evaluation Adaptability of Pasture Grasses to Act as Chemical Filters," Farm Animal Wastes. Symposium May 5–7 1966, Beltsville, Maryland.

54. William McLarney, "The Backyard Fish Farm," *Organic Gardening and Farming's Reader's Research Project No. 1,* 1973.

55. William McLarney, "An Introduction to Aquaculture on the Organic Farm and Homestead," *Organic Gardening and Farming.* August 1971. pp 71–76.

56. William McLarney, "Aquaculture: Toward an Ecological Approach," in *Radical Agriculture,* edited by Richard Merrill (New York: Harper & Row), forthcoming.

57. Fry, "Farmer Turns Pig Manure."

58. Clarence G. Golueke and William J. Oswald, "Biological Conversion of Solar Energy to the Chemical Energy of Methane," *Applied Microbiology* 7(1959):219–227.

59. William J. Oswald and Clarence G. Golueke, "Biological Transformation of Solar Energy," *Advances in Applied Microbiology* 2(1960):223–262.

60. William J. Oswald and Clarence G. Golueke, "Solar Power via a Botanical Process," *Mechanical Engineers.* February 1964. pp 40–43.

61. Clarence C. Golueke and William J. Oswald, "Harvesting and Processing Sewage-Grown Planktonic Algae," *Journal of the Water Pollution Control Federation.* 1965.

Solar Power

1. Steve Baer, "Solar House," *Alternative Sources of Energy* 10 (March 1973):8.

2. "Plowboy Interview with Steve Baer," *The Mother Earth News* 22:6–15.

3. Zomeworks Corporation, PO Box 712, Albuquerque, NM 87103.

4. H. C. Hottel and Duane D. Erway, "Collection of Solar Energy," in *Introduction to the Utilization of Solar Energy,* edited by Zarem and Erway. (New York: McGraw-Hill, 1963.)

5. George O. G. Löf, "The Heating and Cooling of Buildings with Solar Energy," in *Introduction to the Utilization of Solar Energy*, edited by Zarem and Erway (New York: McGraw-Hill, 1963.), out of print.

6. H. E. Thomason, "Experience with Solar Houses," *Solar Energy*. First Quarter 1966, pp. 17–22.

7. H. E. Thomason, "Solar House Models," a booklet which can be purchased from Edmund Scientific Co., 150 Edscorp Building, Barrington, NJ 08077.

8. H. E. Thomason, "Solar House Plans," Similar to reference 7 only it contains house plans and more up-to-date information.

9. A license can be purchased from Edmund Scientific Co. (see reference 7) which will allow you to build one of Thomason's houses.

10. Steve Baer, "Zomeworks," *Alternative Sources of Energy* 6(June 1972):18. A short note on a convective air loop rock storage solar heater.

11. Gar Smith, "Air Conditioning with the Sun and Outer Space," *Alternative Sources of Energy* 10 (March 1973):15.

12. Harold Hay, "A Short Letter on the Pros and Cons of Manual and Automatic Control," *Alternative Sources of Energy* 12 (October 1973):52.

13. News of Iowa State, March/April 1974, Morrill Hall, Iowa State University, Ames, IA 50010.

14. A. B. Makhijani and A. J. Lichtenberg, "Energy and Well-Being," *Environment* 14 (June 1972):10.

15. Roger Douglas, "Available Solar Heat," *Alternative Sources of Energy* 13 (March 1974):19–21.

16. Ken Orvis, "Windowbox," *Alternative Sources of Energy* 13(February 1974):51.

17. J. L. Threlkeld and R. C. Jordan, "Direct Solar Radiation Available on Clear Days," ASHRAE Trans. 64(1958):45–56.

Appendix I

1. For further elucidation on this and most other points in this article, see Donald Marier and Ronald Weintraub, "Local Energy Production for Rural Homesteads and Communities," in *Radical Agriculture*, edited by Richard Merrill, (New York: Harper & Row), forthcoming. Also published as a separate pamphlet by *Alternative Sources of Energy*.

2. E. W. Golding, *The Generation of Electricity by Wind Power*, (New York: Philosophical Library, 1955).

3. R. C. Schlichtig and J. A. Morris, Jr., "Thermoelectric and Mechanical Conversion of Solar Power," *Journal of the Solar Energy Society* 3(April 1959).

4. Don Marier, "Building a Hydraulic Ram," *Alternative Sources of Energy: Practical Technology and Philosophy for a Decentralized Society*.

5. For plans and specifications on building water wheels, see Vic Marks, *Cloudburst: a Handbook of Rural Skills and Technology*, Cloudburst Press, Box 79, Brackendale, BC, Canada, 1973.

6. Phil Carabateas, "More on Alcohol and Wood Gas," *Alternative Sources of Energy* 10(March 1973)21.

7. Golding and Thacker, "The Utilization of Wind and Solar Radiation, and Other Local Energy Resources for the Development of a Community in an Arid or Semi-Arid Area," *Symposium of the Use of Wind Power and Solar Energy in Arid Areas,* New Delhi, India.

8. Several more examples are given in Marier and Weintraub, "Local Energy Production."

9. *Ibid.*

10. Egon Glesinger, *The Coming Age of Wood.* (New York: Simon & Schuster, 1949), out of print.

11. Write U.S. Weather Service, c/o U.S. Weather Bureau, Dept. of Commerce, Washington, DC 20230, or telephone local Weather Bureau.

12. American Society of Heating and Air Conditioning Engineers (ASHRAE), 345 E. 47 St., New York, NY 10017.

13. S. Visher, *Climatic Atlas of the United States.* (Cambridge, MA: Harvard University Press, 1954).

14. For a detailed examination of a community in another region of the U.S.—rural southern Wisconsin—see Marier and Weintraub, "Local Energy Production."

15. T. G. Hicks, *Pump Selection and Application.* (New York: McGraw-Hill, 1957).

16. Marier and Weintraub, "Local Energy Production."

17. J. Dreyfuss, "Dome Home: Heat from Sun, Electricity from Wind," *St. Paul Pioneer Press.* (Sunday, January 21 1973).

18. Bruce Haggart and Graham Caine, "Ramification and Propagations of Street Farm," *Undercurrents in Science and Technology* 4(Spring, 1973).

19. 1973 press release from California State College, California, PA 15419.

20. Richard Blazej, et. al., "Plans for Grassy Brook Village—a prospectus." RFD 1, Newfane, VT 05345, 1973.

BIBLIOGRAPHY

Wind Power

Baumeister, Theodore, and Lionel Marks. *Standard Handbook for Mechanical Engineers.* 7th ed. (New York: McGraw-Hill Book Co., 1967). Contains a 6-page section on wind power, including many formulas and tables.

Benson, Arnold. *Plans for the Construction of a Small Wind Electric Plant.* Publication No. 33. Available from Oklahoma State University, Stillwater, OK 74079.

Brace Research Institute. *How to Construct a Cheap Wind Machine for Pumping Water.* Leaflet No. L–5. 2nd rev. ed. (Quebec: Macdonald College of McGill University, 1965). Furnishes instructions and diagrams for units "suitable for home construction by the handyman." Available from Brace Research Institute, Macdonald College of McGill University, Ste. Anne de Bellevue 800, Quebec, Canada.

———. *Notes on the Development of the Brace Airscrew Windmill as a Prime Mover.* Leaflet no. R–38, 1969. Available from the above address.

———. *A Simple Electric Transmission System for a Free Running Windmill.* Leaflet No. T–68, 1970. Available from above address.

Clark, Wilson. "Interest in Wind is Picking Up as Fuels Dwindle," *Smithsonian.* November 1973, p. 70. A good, non-technical overview.

Climatic Atlas of the United States. Gives wind averages, the strongest wind, and wind direction. The atlas and a monthly wind report for each state are available from Environmental Data Service, National Climatic Center, Federal Bldg., Asheville, NC 28801.

"Device Gauges Wind for the Amateur Weatherman," *Popular Science Monthly.* October 1948, p. 236.

Fales, E. N. "Windmills," *Mechanical Engineers' Handbook.* 2nd rev. ed. (New York: McGraw-Hill, pp. 1080–85).

Fasching, Lanny. "Whirlwind," *Mechanix Illustrated.* April 1941, pp. 102–3. Plans for a small wind generator.

Frankiel, J. "Wind Power Research in Israel," *Wind and Solar Energy.* UNESCO 1956. Evaluates some small wind generators.

Freese, Stanley. *Windmills and Millwrighting* (New York: Cambridge University Press, 1957). Deals with old windmills.

"Giant Wind Machine for Generating Electricity Gets Federal Scrutiny," *National Observer.* June 24 1972. A short article on a wind generator that would run a circular track, employing an airfoil for propulsion. The device was designed by Fred Davison, a rancher from Highwood, MT.

Glavert, H. *Aerodynamic Theory*, Volume 4 (New York: Dover Publications, 1963). Contains a chapter on windmills and fans.

Golding, E. W. "The Economic Utilization of Wind Energy in Arid Areas," *Wind and Solar Energy*. UNESCO 1956.

———. *The Generation of Electricity by Wind Power* (New York: Philosophical Library, 1956). A very informative work on wind power, with an extensive bibliography.

———. "Wind–Generated Electricity and Its Possible Use on the Farm," *Farm Mechanization*. March 1953.

Grafstein, Paul and Otto B. Schwarz. *Pictorial Handbook of Technical Devices* (New York: Chemical Publishing Co., 1971). A reference that supersedes *The Engineers' Illustrated Thesaurus* (by the same publisher and now out of print). The book's illustrations should assist the designer or builder in visualizing his or her creation, whether conceptive or actual.

Hewson, E. Wendell. *Wind Power*. First Progress Report on Research on Wind Power Potential in Selected Areas of Oregon, Report PUD 73–1, 1973. Discusses windmill sites in the Oregon area, explains the use of topographic models placed in wind tunnels to study local wind movement, and contains an extensive bibliography. Available from Oregon P.U.D. Directors' Assoc., c/o Central Lincoln P.U.D., 255 SW Coast Highway, Newport, OR 97365.

Hutter, U. "Planning and Balancing of Energy of Small Output Wind Power–Plant," *Wind and Solar Energy*. UNESCO 1956.

Juul, J. "Wind Machines," *Wind and Solar Energy*. UNESCO 1956. A very detailed paper on Danish wind generator technology.

McCaull, Julian. "Windmills," *Environment*. January/February 1973, pp. 6–17. A description of William Heronemus' plan for large-scale power generation in the Great Plains. The author discusses the feasibility of large-scale projects, citing the Grandpa's Knob experiment and recent research done outside the U.S.

McKinley, James L., and Ralph D. Bent. *Powerplants for Aerospace Vehicles*. 3rd rev. ed. (New York: McGraw-Hill Book Co., 1965). A reference for information on the balancing of high-speed windmill blades. Previously published as *Aircraft Power Plants*.

Manikowske, Wallace. *Windmill Electric Lighting and Power*. Bulletin No. 105. North Dakota Agricultural Experiment Station, 1915. A thesis for a Bachelor's Degree, describing various propellers and generation systems.

Meyer, Hans. "Wind Energy," *Domebook 2*. p. 121. Available from Pacific Domes, Box 279, Bolinas, CA 94924.

———. "Wind Generators: Here's an Advanced Design You Can Build," *Popular Science Monthly*. November 1972, pp. 103–5. Directions and plans for constructing a light-duty home unit, using expanded paper for rotor blades.

Pigrand, F. D., and Rex Weiles. "Power in the Wind," *New Scientist*. May 1965. A description of a Dutch windmill used to generate electricity.

"Plowboy Interview with Marcellus Jacobs," *Mother Earth News*. 24: 52.

Putnam, Palmer C. *Power from the Wind* (New York: Van Nostrand Co., 1948). This out-of-print work gives a good description of the Grandpa's Knob experiment, has a good bibliography, and includes sections on wind data analysis, theory of wind power, and site selection.

RedRocker, Winnie. "Build a Wind Generator!" *Lifestyle!* 3: 47–51. Describes how to construct a two-bladed wooden rotor. This article was originally published in *Alternative Sources of Energy*, No. 8.

Sencenbaugh, Jim. "I Built a Wind Charger for $400.00!" *Mother Earth News* 20: 32. Description of a home-built wind generator design.

Stokhuyzen, Frederick. *The Dutch Windmill* (New York: Universe Books, 1967).

United Nations Conference on New Sources of Energy. *Proceedings, Vol. 7, Wind Power* (New York: United Nations Press, 1964). Includes 40 articles on wind power, wind behavior, site selection, home-built designs, and recent developments. Out of print.

Voigt, Helmut. *Principles of Steel Construction Engineering in the Building and Operation of Wind-Driven Power Plants.* The author claims that one 30-meter wind power plant is capable of supplying a small town with all of its light, power, and heat requirements. Available from National Technical Information Service, U.S. Dept. of Commerce, Springfield, VA 22150.

Volunteers in Technical Assistance (VITA) offers several plans for windmills. A catalog of "Village Technology Plans" is available from VITA, 3706 Rhode Island Ave., Mt. Ranier, MD 20822.

Wailes, Rex. *The English Windmill* (New York: Augustus M. Kelley, 1967). A historic view of windmills, with details of construction.

Wind Energy Bibliography. A 64-page booklet, available from Windworks, Box 329, Rte. 3, Mukwonago, WI 53149.

"Windpower," *Sierra Club Bulletin.* September 1971.

World Meteorological Organization. *Energy from the Wind.* Geneva, 1954. An assessment of suitable winds and sites, with an 18-page bibliography on wind energy and its utilization.

Water Power

Anderson, A. P. *Domestic Water Supply and Sewage Disposal Guide* (New York: Theodore Audel & Co., 1967). Includes a chapter on hydraulic rams.

Brown, J. Guthrie, ed. *Hydro Electric Engineering Practice* (New York: Gordon & Breach, 1958; London: Blackie & Sons, Ltd., 1958). A very complete treatise, covering the entire field of hydroelectric engineering.

Cetin, Frank. "When Water Turned the Wheels," *Wisconsin Tales and Trails.* April 1968.

Creager, W. P., and J. D. Justin. *Hydro Electric Handbook.* 2nd ed. (New York: John Wiley and Son, 1950). A comprehensive handbook, especially good for reference.

Crowley, C. A. "Power from Small Streams," Part I. *Popular Mechanics.* September 1940, p. 466.

———. "Power from Small Streams," Part II. *Popular Mechanics.* October 1940, pp. 626–30.

Davis, Calvin V. *Handbook of Applied Hydraulics.* 2nd ed. (New York: McGraw-

Hill Book Co., 1952). A comprehensive handbook, covering all phases of applied hydraulics. Several chapters are devoted to hydroelectric application.

Haimerl, L. A. "The Cross Flow Turbine," *Water Power*. January 1960. This article describes a type of water turbine which is being used extensively in small power stations, especially in Germany. Reprints are available from Ossberger Turbinenfabrik, 8832 Weissenburg, Bayern, Germany.

How to Install Plastic Pipe for Drain, Waste, and Vent Lines.

How to Install Plastic Pipe for Hot and Cold Water Lines. These pamphlets are available from Sears Catalog Order Plant, Chicago, IL 60607.

Intermediate Storage for Farmstead Water Systems. Available from Cooperative Extension Service, University of Maryland, College Park, MD 20740.

Logsdon, Gene. *Homesteading* (Emmaus, PA: Rodale Press, 1973). Describes methods of using water power for the homesteader, and tells how to compute the potential of a stream, pp. 226–29.

Marier, Don. "Hydraulic Rams," *Alternative Sources of Energy* 1:3–7.

Mockmore, C. A., and F. Merryfield. *The Banki Water Turbine.* Oregon State College Engineering Experiment Station Bulletin No. 25, Corvallis, Oregon February 1949. This translation of a paper given by Donat Banki gives a highly technical description of the Banki turbine, together with the results of tests.

Old Mill News. Available from Society for the Preservation of Old Mills, Box 435, Wiscasset, ME 04578. The Society acts as a clearing house for information on old water mills and related subjects.

"The Owner-Built Hydroelectric Plant," *Access Catalog* 1 (7): 12–15. Information on measuring water flow, building dams, and types of waterwheels.

Paton, T. A. L. *Power from Water* (London: Leonard Hill, 1961). A concise general survey of hydroelectric practice.

Plumbing for the Home and Farmstead. Farmer's Bulletin No. 2213. Available from the U.S. Government Printing Office, Washington D.C. 20401.

Volunteers in Technical Assistance (VITA) has plans for a 1-kw river generator. A catalog of "Village Technology Plans" is available from VITA, 3706 Rhode Island Ave., Mt. Ranier, MD 20822.

Water Supply Sources for the Farmstead and Rural Home. Farmer's Bulletin No. 2237. Available from the U.S. Government Printing Office, Washington D.C. 20401.

Zerban, A. H., and E. P. Nye. *Power Plants.* 2nd ed. (Scranton: International Text Book Co., 1952). Chapter 12 gives a concise presentation of hydraulic power plants.

Wood Power

Bacon, R. M. "Managing the Small Woodlot," *Yankee*. January 1974, p. 156. A general, non-technical article. The author refers those desiring specific information to the local county forestry agent or state university.

Cooperative Extension, New York State College of Agriculture and Life Sciences. The college publishes several Information Bulletins on planting and managing a farm woodlot. Their catalog is available from Mailing Room, Bldg. 7, Research Park, Cornell University, Ithaca, NY 14850.

Cox, Jeff. "Heating Your Home Without Harming Nature," *Organic Gardening and Farming*. March 1973, pp. 90–93. Gives information on calculating the energy yield of various woods.

Emery, Carla. "Of Wood Cook Stoves," *Organic Gardening and Farming*. October 1973, pp. 48–51.

"The Energy Crisis Vs. Wood Fuel," *Conservation Circular* 11 (Summer 1973) 3. Discusses the suitability of different woods as fuel, surveys various types of wood burning units on the market, and compares the energy value of various woods with coal, fuel oil, and natural gas.

Glesinger, Egon. *The Coming Age of Wood* (New York: Simon & Schuster, 1949). Out of print.

Jordan, Charles J., and Jessie S. Cole. "The Shape of Things to Come" *Yankee*. January 1974, p. 122. An interesting article on the use of wood stoves in New England that describes the recent revival of wood as a heating fuel.

Managing the Family Forest. Farmer's Bulletin No. 2187. U.S. Dept. of Agriculture, 1962. Covers restocking, thinning, harvesting, marketing, etc. Available from The Superintendent of Documents, U.S. Government Printing Office, Washington D.C. 20402.

Rolerson, Darrell A. "The Woodlot: A Balance in the Ecology of Your Farm" *Organic Gardening and Farming*. March 1973, p. 94. The experiences of a Maine homesteader, including a section on "What to Look for in a Chain Saw."

Volunteers in Technical Assistance (VITA). *Wood Baking Stove*. Plans and instructions, available from VITA, 3706 Rhode Island Ave., Mt. Rainier, MD 20822.

Willey, E. C. *Rating and Care of Domestic Sawdust Burners*. Bulletin Series, No. 15. (Corvallis: Oregon State College, 1941). A technical study, available from Oregon State Engineering Experiment Station, Corvallis, OR 97331.

Wood Fuel Preparation. U.S. Forest Service Research Note FPL–090. U.S. Dept. of Agriculture, 1965. Information on the preparation of fuel from wood residues and other wood raw materials. Available from Forest Products Laboratory, Forest Service, U.S. Dept. of Agriculture, Madison, WI 53705.

Methane

Auerback, Les, William Olkowski, and Ben Katz. *A Homesite Power Unit: Methane Generator*. A good introduction to methane generation. This booklet gives an account of a generator designed and constructed at the University of California at Berkeley, and includes information on building and operating a plant. Available from Les Auerbach, 242 Copse Rd., Madison, CT 06443.

Bohn, Hinrich L. "A Clean New Gas," *Environment* 10 (1971):49.

Compost Science. This bimonthly journal has articles on bio-gas plants and anaerobic digestion of waste. Published by Rodale Press, Inc., Emmaus, PA 18049.

Fry, L. J. "Power and Electric Light from Pig Manure," *Farm & Country*. April 1960, pp. 428–29.

Grout, A. Roger. *Methane Gas Generation from Manure.* A pamphlet giving details of the methane generators constructed for the 1973 Pennsylvania Agricultural Progress Days at Hershey. Available from Dept. of Agriculture, Commonwealth of Pennsylvania, 2301 N. Cameron St., Harrisburg, PA 17120.

Hutchinson, T. H. "Methane Farming in Kenya," *Compost Science.* November/-December 1972, pp. 30–31.

Klein, S. A. "Methane Gas—An Overlooked Energy Source," *Organic Gardening and Farming.* June 1972, pp. 98–101.

Methane Systems. This 27-page booklet gives the theory and design for a two-stage, continuous-feed generator. Available from Earthmind, 26510 Josel Dr., Saugus, CA 91350.

Mother Earth News Reprints. No. 102, "The Plowboy Interview: Ram Bux Singh." No. 144, two articles, one with background information and the other with diagrams and technical advice. No. 162, "The Plowboy Interview: L. John Fry." All available from *The Mother Earth News* at PO Box 70, Hendersonville, NC 28739.

Singh, Ram Bux. *Bio-Gas Plant—Generating Methane from Organic Wastes,* 1971. A thorough treatment of the subject, detailing plant design and construction and describing how a plant works. Available from James Whitehurst, Ral-Jim Farm, Benson, VT 05731.

———. "Building a Bio-Gas Plant," *Compost Science.* March/April 1972, pp. 12–16. Gives detailed instructions for five designs, some simple and others more sophisticated.

———. "Generating Methane from Organic Wastes," *Compost Science.* January/February 1972, pp. 20–25. Gives the details of designing a bio-gas plant.

Taiganides, Eliseos P. "Anaerobic Digestion of Poultry Wastes," *World Poultry Science Journal.* October/December 1963, pp. 252–62.

———. "Manure Gas Plants," *National Hog Farmer.* Swine Information Service, Bulletin no. F 13, May 1963.

Volunteers in Technical Assistance (VITA) has plans for a methane generator. A catalog of "Village Technology Plans" is available from VITA, 3706 Rhode Island Ave., Mt. Ranier, MD 20822.

Solar Power

Altman, Manfred. *Conversion and Better Utilization of Electric Power by Means of Thermal Energy Storage and Solar Heating.* Report UPTES–71–1. Gives information on storage of heat energy. Available from U.S. Dept. of Commerce, NTIS, 5285 Port Royal Rd., Springfield, VA 22151.

American Society of Heating, Refrigeration and Air Conditioning Engineers (ASHRAE). *Low Temperature Engineering Application of Solar Energy.* ASHRAE Technical Committee on Solar Energy Utilization, 1967. Available from ASHRAE, 345 E. 47 St., New York, NY 10017.

Anderson, Bruce. *Solar Energy and Shelter Design.* 1973. A comprehensive work on solar heating with a 12-page bibliography. Available from the author at Box 47, Harrisville, NH 03450.

Ayres, E. "Power from the Sun," *Scientific American.* July 1950, pp. 16–21.

Boer, K. W. "The Solar House and Its Portent," *CHEMTECH.* July 1973, pp. 394–400. Discusses the practicality of the solar house and analyzes the cost of a solar heating system.

Brace Research Institute. *How to Build a Solar Water Heater.* Leaflet No. L–4. 2nd rev. ed. Quebec: Macdonald College of McGill University, 1973. Plans for a solar water heater, suitable for domestic or agricultural use in sunny areas. Available from Brace Research Institute, Macdonald College of McGill University, Ste. Anne de Bellevue 800, Quebec, Canada.

———. *Solar Steam Cooker.* Leaflet No. L–2. 2nd rev. ed. 1972. Plans for a cooker for boiling foods in sunny areas. Available from the above address.

———. *How to Heat Your Swimming Pool Using Solar Energy.* Leaflet No. L–3. 1965. Available from the above address.

Branley, Franklin M. *Solar Energy.* (New York: Thomas Y. Cromwell Co., 1957).

Brinkworth, B. J. *Solar Energy for Man.* (New York: John Wiley & Sons). A good introduction to the whole field of solar energy, addressed to the general reader.

Chinnery, D. N. W. *Solar Water Heating in South Africa.* National Building Research Council Bulletin 44. Pretoria, South Africa: Council for Scientific and Industrial Research, 1967.

Clark, Wilson. "How to Harness Sun-Power and Avoid Pollution," *Smithsonian.* November 1971, pp. 14–21.

Daniels, Farrington. *Direct Use of the Sun's Energy.* (New Haven: Yale University Press, 1964). Good background reading. Out of print, but available in paperback from Ballantine Books, New York.

———. "The Solar Era: Part II—Power Production with Small Solar Engines," *Mechanical Engineering.* September 1972.

Daniels, Farrington, and John A. Duffy. *Solar Energy Research.* (Madison: University of Wisconsin Press, 1955).

Deitz. "Large Enclosures and Solar Energy," *Architectural Design.* April 1971.

Farber, E. A. "Solar Air-Conditioning System," *Solar Energy Journal.* No. 291, 1966.

Feasibility of Solar Energy as a Major Power Source. Citizens' Organization for the Study of Solar Energy, 1971. Available from the Organization at Rte. 8, Box 550–B, Tucson, AZ 85710.

Halacy, D. S., Jr. *The Coming Age of Solar Energy.* 2nd rev. ed. (New York: Harper & Row, 1973).

Hamilton, R. W., ed. *Space Heating with Solar Energy* (Cambridge: Massachusetts Institute of Technology, 1954).

Hammond, Allen L. "Solar Energy: The Largest Resource," *Science.* September 22 1972, pp. 1088–90. A brief survey of solar energy research.

Holloway, D. G. *The Physical Properties of Glass* (New York: Springer-Verlag, 1973). Although this book is heavy reading, it is a valuable reference because of the importance of glass in solar energy collection.

"The House that Stores the Sun," *Popular Mechanics.* October 1957, p. 158.

Institute of Solar Energy Conversion. *Reference List of Major Publications on Solar*

Energy. Available from the Institute at University of Delaware, Newark, DE 19711.

Löf, G. O. G., D. A. Fester and J. A. Duffy. "Energy Balances on a Parabolic Cylindrical Collector," *Transactions of the American Society of Mechanical Engineers*. 1962, pp. 24–32.

Low Temperature Engineering Application of Solar Energy. Chapters cover availability and measurement of solar radiation, design factors, selective surfaces, potential uses in the tropics, and various types of solar water heaters. The book is available from ASHRAE, 345 E. 47 St., New York, NY 10017.

Mathur, Khanna, et. al. "Domestic Solar Water Heater," *Journal of Scientific Industrial Research*. February 1959, pp. 15–58.

Morgan, Scott and Chole; and David and Susan Taylor. *Hot Water*. 1974. This booklet gives detailed directions for the construction of two home heating systems, one solar powered, and the other employing a stack coil to take heat from a chimney. Available from *Hot Water*, 350 E. Mountain Dr., Santa Barbara, CA 93108.

Morrow, Walter E., Jr. "Solar Energy: Its Time is Near," *Technology Review*. December 1973, pp. 31–43. Surveys solar energy research and gives an excellent evaluation of the practicality of solar energy systems, both large and small.

"Portable Tank Captures Solar Energy," *Plastic World*. March 1972, p. 32.

Proceedings of the World Symposium on Applied Solar Energy, Phoenix, Arizona. (Menlo Park: Stanford Research Institute, 1956).

Progress Report No. 2. Describes the project to harness the sun with trough-type paraboloids that concentrate sunlight on heat pipes, conducted by the University of Minnesota and Honeywell. Available from Prof. R. C. Jordan, Dept. of Mechanical Engineering, University of Minnesota, Minneapolis, MN 55455.

Rau, Hans. *Solar Energy* (New York: MacMillan Co., 1964).

Schlichtig, R.C., and J. A. Morris, Jr. "Thermoelectric and Mechanical Conversion of Solar Power," *Journal of the Solar Energy Society* 2 (1959).

Salam and Daniels. *Selective Radiation Coatings for Solar Heating* (Madison: University of Wisconsin, 1958). A booklet on the preparation and testing of coatings.

O'Connor, Egan. "Solar Power for the Seventies," *High Country News*. January 4 1974, p. 1. Gives a brief overview, along with photos and description of solar heating systems developed by Zomeworks. The *News* is an environmental biweekly published in Lander, WY.

Selected Bibliography on Solar Energy Conversion, 1972. Available from Optical Sciences Center, University of Arizona, Tucson, AZ 85721.

Solar Energy Applications. Available from Aurthur D. Little, Inc., 20 Acorn Park, Cambridge, MA 02140.

Solar Energy as a National Resource. NSF/NASA Solar Energy Panel Report, 1972. Available from Solar Energy Panel, Dept. of Mechanical Engineering, University of Maryland, College Park, MD 20742.

Solar Energy Digest. PO Box 17776, San Diego, CA 92117. Published monthly, the *Digest* is an excellent source of continuing work on solar energy.

Solar Energy—House Heating. A bibliography prepared by the Library of Congress. Available from L.C. Tracer Bullet, Reference Section, Science & Technology Div., Library of Congress, 10 First St., SE, Washington, D.C. 20540.

Solar Energy Industry Report and *Solar Energy Washington Report.* Subscriptions to these biweekly newsletters may be expensive for the individual. Available from The Solar Center, 1001 Connecticut Ave., NW, Washington, D.C. 20036.

Solar Energy: The Journal of Solar Energy Science and Technology. Published by the International Solar Energy Society, c/o Smithsonian Radiation Biology Laboratory, 12441 Parklawn Dr., Rockville, MD 20852.

"Solar Energy Technology: New Seriousness," *Science News.* April 8 1972, pp. 225–40.

Thin Films and Solar Energy. This booklet describes the uses of antireflection coatings for flat plate collectors. It is available from Optical Coating Laboratory, Inc., PO Box 1599, Santa Rosa, CA 95403.

Thomason, Harry E. "House with Sunshine in the Basement," *Popular Science.* February 1965, pp. 89–92.

Thomsen, Dietrick E. "Farming the Sun's Energy," *Science News.* April 8 1972, pp. 237–38.

Transaction of the Conference on the Use of Solar Energy: The Scientific Basis (Tucson: University of Arizona Press, 1958).

UNESCO Features, No. 656 (1973). An issue on the uses of solar energy, including: an interview with Prof. Felix Trombe, a French expert on solar energy; an article describing a French solar home; and information on solar pumps and distillation plants.

U.S. House of Representatives. *Solar Energy for Heating and Cooling.* Hearings before the Subcommittee on Energy of the Committee on Science and Astronautics: 93rd Congress, 1st Session, No. 13, 1973. A 291-page compilation of committee transcriptions and reports from experts in the field of solar energy.

U.S. House of Representatives. *Solar Energy for the Terrestrial Generation of Electricity.* Hearings before the Subcommittee on Energy of the Committee on Science and Astronautics: 93rd Congress, 1st Session, No. 12, 1973. Includes Dr. Meinel's report on "Progress in Solar Photothermal Power Conversion," and a report of Honeywell's research with the large-scale generation of electric power from solar energy.

U.S. House of Representatives. *Solar Energy Research—A Multidisciplinary Approach.* Staff Report of the Committee on Science and Aeronautics. A report on the extent of Federal and private research currently being conducted and on expected future levels of funding.

Weingart, Jerome. "Everything You've Always Wanted to Know About Solar Energy, but Were Never Charged Up Enough to Ask," *Environmental Quality Magazine.* December 1972, pp. 39–43. A good introduction, touching on solar water heating, space heating and cooling, and electric power conversion with thermal collectors, solar cells, and microwave-transmitting space stations.

Williams, D. A., et. al. "Intermittent Absorption Cooling Systems with Solar Regeneration," *Refrigeration Engineering.* November 1958.

Zarem and Erway, eds. *Introduction to the Utilization of Solar Energy.* University of California Engineering and Sciences, Extension Series. (New York: McGraw-Hill Book Co., 1963).

General

Alternative Sources of Energy. Don Marier, Rte. 2, Box 90–A, Milaca, MN 56353. An excellent source of current information, published quarterly. Some back issues are available; also a 280–page book, *Alternative Sources of Energy: Practical Technology & Philosophy For a Decentralized Society.*

American Association of Physics Teachers. Resource Letter ERPEE–1 on Energy: Resources, Production and Environmental Effects. A list of published references and organizations, available from Executive Officer, American Assoc. of Physics Teachers, 1785 Massachusetts Ave., NW, Washington, D.C. 20036.

Brace Research Institute. *Publications List.* Miscellaneous Report No. M–17 (Quebec: Macdonald College of McGill University, 1973). A list of reprints and reports that are available from the Institute. Most titles are concerned with helping the individual or small community to utilize local resources, whether human, material, or energy. Availble from Brace Research Institute, Macdonald College of McGill University, Ste. Anne de Bellevue 800, Quebec, Canada.

Burstall, Aubrey F. *Simple Working Models of Historic Machines* (Cambridge: Massachusetts Institute of Technology Press). Instructions for making such models as a sun-tracking clockwork, a hydraulic ram, and a variety of pumps.

CF Letter. A newsletter reporting on environmental issues, published by the Conservation Foundation, 1717 Massachusetts Ave., Washington, D.C. 20036.

Clarke, Robin, and John Todd. "The Third Alternative," *Harper's Magazine.* In press. An interesting, humanitarian discussion of an alternative technology—one that would use technological advances to permit self-sufficiency on a decentralized level.

Crouch, Gerrard A. *The Autonomous Servicing of Dwellings—Design Proposals* (Cambridge: University of Cambridge, 1972). A booklet that details several ideas for a self-contained solar house, incorporating windmills for electricity and methane generators. Available from University of Cambridge, Dept. of Architecture, Technical Research Div., 1 Scroope Terrace, Cambridge CB2 1 PX, England.

Dreyfuss, John. "Unique Dome Home Harnesses Sun and Wind," *Los Angeles Times.* January 1 1973, Part II, p. 1. A good description of Bob Reines' experiments with construction, solar power, and wind power.

Dubin, Fred S. *Energy Conservation through Building Design and a Wiser Use of Electricity.* This pamphlet is available from Dubin, Mindell, Bloome Associates, 42 W. 39 St., New York, NY 10018.

Eveready Battery Applications Engineering Data, 1971. A good description of all commonly available battery types, along with complete application data. The booklet is available from Union Carbide Corp., Consumer Products Div., 270 Park Ave., New York, NY 10017.

Heronemus, William. "Alternatives to Nuclear Energy," *Catalyst* 2(3):21–26.

Gates, David M. "The Flow of Energy in the Biosphere," *Scientific American.* September 1971, pp. 89–100. An analysis of solar radiation and how it is used by living things.

Hubbert, M. King. "The Energy Resources of the Earth," *Scientific American.* September 1971, pp. 61–70. Describes the flow of energy to and from the earth, gives statistics on the earth's supplies of energy resources, and posits that a shortage of energy must soon cause rates of industrial and population growth to stabilize.

Lyerla, Jim. *Ideas to Help the Meek Inherit the Earth.* Information on windmills, solar heating, and other energy sources. Available from the author at 3228 Keats St., San Diego, CA 92106.

Mother Earth News. A bimonthly magazine with many articles on producing one's own power, available from PO Box 70, Hendersonville, NC 28739.

Proceedings of the United Nations Conference on New Sources of Energy. Volumes 4 through 6 deal with solar energy, and Volume 7 is devoted to wind power. Available from Sales Section, United Nations, New York, NY 10017.

Reynolds, John. *Windmills and Watermills* (New York: Praeger Publishers, 1970). Simple directions for construction, written for the competent mechanic. The mills described are of traditional design.

Roberts, Rex. *Your Engineered House* (New York: M. Evans & Co., 1964).

"Robinson Crusoe in London," *Mechanical Engineering.* October 1973, pp. 46–47. A short description of Graham Caine's "eco-house" in London.

Smith, Gerry E. *Economics of Solar Collectors, Heat Pumps, and Wind Generators.* 1973. A detailed technical paper that investigates the economic feasibility of heating a home with a solar collector and a heat pump driven by a wind-powered generator. Available from University of Cambridge, Dept. of Architecture, Technical Research Div., 1 Scroope Terrace, Cambridge CB2 1 PX, England.

"Solar and Wind Energy," *Research.* March 1961, pp. 82–87.

Spanides, A. G., and Athan D. Hartzikakidis, eds. *Solar and Aeolian Energy* (New York: Plenum Publishing Corp., 1964). Proceedings of the International Seminar On Solar and Aeolian Energy, Sounion, Greece, 1961.

Tamplin, Arthur R. "Solar Energy," *Environment.* June 1973, p. 16. A discussion of the feasibility of solar and methane energy as alternative energy sources.

Conservation

Center for Environmental Studies. *Energy Conservation in Housing: First Year Progress Report.* Report No. 6. (Princeton: Princeton University, 1973). A report on a project to monitor the energy utilization of a residential community. Specifically, the Center examined the development process of this community—from its conception, through construction, and finally to occupancy—with the aim of correlating variables at each stage with energy use. Available from the Center at Engineering Quadrangle, Princeton University, Princeton, NJ 08540.

Citizens' Advisory Committee on Environmental Quality. *Citizen Action Guide to Energy Conservation.* 1973. This booklet details ways in which citizens can reduce the waste of energy, through individual and collective action, both in the home and in local groups. A chapter describes work being done by the government and industry, and includes a list of state contacts for information on energy. Available from Superintendent of Documents, U.S. Government Printing Office, Washington, D.C. 20402.

11 Ways to Reduce Energy Consumption and Increase Comfort in Household

Cooling. Office of Consumer Affairs, 1972. Suggestions for the household consumer, available from the Superintendent of Documents (see above).

McGlennon, John A. S. "Reducing Fuel Consumption in Household Heating," *Environment News* 9:12–13. Suggestions on ways to conserve household and transportation energy.

Rodale, Robert. "The Soil and Energy," *Environment Action Bulletin.* February 24 1973, p. 4. This article contends that agriculture, as practiced in this country today, is wasteful of energy—that is, the yield of food energy is not justified by the expense of energy required to make chemical fertilizers and pesticides and to propell goods to market.

Rubin, Milton D. "Toward A Rational Use of Energy," *Alternative Sources of Energy* 7:12–13. An interesting comparison of the efficiency of "use end devices" in the home. Rubin suggests that more efficient energy-using devices could serve as a short-term means of avoiding an energy crisis, and mentions the heat pump as an example.

7 Ways to Reduce Fuel Consumption in Household Heating. Office of Consumer Affairs, 1972. Suggestions for the household consumer, available from Superintendent of Documents, U.S. Government Printing Office, Washington, D.C. 20402.

Stanford Research Institute. *Patterns of Energy Consumption in the United States* (Menlo Park: Stanford Research Institute, 1972).

GLOSSARY

absorptance—the ratio between the radiation absorbed by a surface and the total energy falling on that surface. A matte black surface, such as used in solar collectors, has a high absorptance.

aerobic bacteria—micro-organisms that require oxygen for life.

alcohol—see ethyl alcohol, methyl alcohol.

alternating current (AC)—electric current which changes its direction of flow at regular intervals, normally making 60 cycles per second. AC is easier to transmit than direct current and is also more easily changed to higher or lower voltages. Household current is AC.

ampere (amp)—the unit of rate of flow in an electric circuit.

ampere-hour—unit of electrical charge, equalling the quantity of electricity flowing in 1 hour past any point of a circuit carrying a current of 1 ampere. Storage batteries are rated in ampere-hours to show the quantity of electricity that can be used without discharging the battery beyond safe limits.

anaerobic bacteria (anerobes)—micro-organisms that can live without the presence of oxygen.

angle iron—an iron or steel bar with an L-shaped cross section.

angle of incidence—the angle at which radiant energy strikes a surface, measuring from the path of the energy to a line perpendicular to that surface at the point of impact.

British thermal unit (Btu)—a unit used to measure quantity of heat; technically, the quantity of heat required to raise the temperature of 1 pound of water 1° F. One Btu=252 calories.

calorie—a unit used to measure quantity of heat; technically, the quantity of heat required to raise 1 gram of water from 14.5° to 15.5°C.

centrifugal pump—a high speed pump that drives water with a rotating impeller.

combined energy system—a plan that utilizes several sources of energy in the most efficient way.

conifers—species of trees which usually keep their leaves in the autumn, including the evergreens.

convected heat—heat which is transferred from one position to another, driven by the change in a gas's density that accompanies a change in temperature.

cord—a unit of volume measurement, 4-by-4-by-8 feet, generally used to measure quantities of wood cut for fuel.

creosote—an oily, odorous distillate of wood tar that may collect on the walls of a chimney as a result of incomplete combustion.

deciduous—species of trees which shed their leaves in the autumn.

demand time—occurs when energy is needed, as heat is at night.

diffuse sky radiation—light that has travelled an indirect path from the sun (that is, light on a cloudy day and light from an obscured sun), as opposed to direct beam sunlight.

direct beam sunlight (incident energy, incident sunlight)—light that has travelled a straight path from the sun, as opposed to diffuse sky radiation.

direct current (DC)—electric current which flows in one direction. Generators produce DC current.

efficiency—a measure of how much of the energy applied to a device is utilized in useful work. For example: if an electric motor is rated 80 percent efficient, then 80 percent of the electrical energy is converted into mechanical motion and the remaining 20 percent is dissipated as heat.

energy conversion—the changing of one form of energy (often in a non-usable state) into another form of energy (usually more useful). For example: converting solar energy to mechanical energy (generation of steam in a solar collector) to electrical energy (by means of a turbine-powered generator). A *direct* energy conversion is one requiring a single change of state.

electrolysis—the process of decomposing a compound by passing an electric current through it.

energy—the ability to do work. Units of energy are: kilowatt-hours (kwh); British thermal units (Btu); and horsepower-hour (hph).

ethyl alcohol (ethanol, grain alcohol)—a colorless liquid of pleasant odor that burns with a pale blue, transparent flame to produce water plus carbon dioxide. The vapor forms an explosive mixture with air and can be used to fuel internal combustion engines. It is most easily produced by the fermentation of carbohydrates.

eutectic salt—characterized by melting and solidifying over a slight range of temperature, a property that makes this substance useful for heat storage. Glauber salt has this eutectic property.

fermentation—chemical changes caused in materials containing sugar or starch by the action of enzymes produced by living organisms.

friable—easily crumbled or reduced to powder.

Glauber salt—see eutectic salt.

head—a body of water kept in reserve at a height, as in a dam; the vertical distance of that level above another point; the resulting pressure of that water at some lower level, as at the turbine or water wheel of a dam.

horsepower—a measure of the rate of doing work, equal to 33,000 foot pounds or 754.2 watts.

humus—organic matter (animal and plant) in a state of decomposition, forming an essential element of all fertile soils.

incidence—the falling of a light ray on a surface.

incident energy (incident sunlight)—see direct beam sunlight.

inverter—a device for converting direct current (DC) into alternating current (AC).

kilowatt—a unit of power equal to 1,000 watts.

methane—an odorless, colorless gas, nearly insoluble in water, burning with a pale, faintly luminous flame to produce water and carbon dioxide (or carbon monoxide if oxygen is defficient). It is the product of the anaerobic fermentation of organic matter.

methyl alcohol (methanol, wood alcohol)—a colorless liquid of pleasant odor, burning with a pale blue, transparent flame to produce water and carbon dioxide. It may be produced by the destructive distillation of hardwood. Methyl alcohol is used as fuel for the internal combustion engines of some racing cars.

monoculture—growing a single crop on a given piece of land.

penstock—the water conduit in a dam that connects the intake to the turbine or water wheel.

pH—the expression of acidity or alkalinity of a given substance. A value of 0 to 7 indicates acidity, and a value of 7 to 14 indicates alkalinity, while 7 is regarded as neutral.

power—the rate of doing work. Units of power are: kilowatts (kw); British thermal units per second; horsepower (hp); and foot-pounds per second, or erg-seconds.

ohm—the unit of resistance in an electric circuit.

rectifier—a device that converts alternating current (AC) into direct current (DC).

sky light—see diffuse sky radiation.

slip ring commutator—a device for periodically changing the direction of flow of electric current by means of two rings that alternately pick up current.

supply time—occurs when energy is supplied in its most convenient form, such as solar radiation during sunny days.

time requirements—the various uses of energy in a household can be classified according to time requirements: a *precise* time requirement, such as lighting, occurs at a certain time of day; a *random* time requirement, such as pumping water to a storage tank, can be fulfilled whenever there is an excess of energy.

torque—the movement of a force around an axis (rotation).

venturi effect—the increase in velocity of a flow of fluid which is created by a constriction in a tube, caused by the fact that the velocity of flow of a fluid is inversely proportional to the pressure in the constricted area.

volt—the unit of pressure in an electric circuit.

voltage regulator—a device used to maintain constant generator voltage.

watt—the unit of rate at which work is done in an electrical circuit, equal to the rate of flow (amperes) multiplied by the pressure of that flow (volts).

zeolite (boiling stone)—a porous mineral that loosely holds water molecules. When heated, the stone melts and gives off water.

THE CONTRIBUTORS

Henry Clews, an aeronautical engineer and a former high school science teacher, lives with his family on their 50-acre homestead in East Holden, Maine. All of the Clews' electrical power for running their hi-fi, television, workshop tools, water pump, and lights comes from the two wind generators—an Australian Dunlite and a Swiss-made Elektro—that they installed on their farm. Henry is so enthusiastic about the potentials of wind power that he and his family have formed the Solar Wind Company, one of the few commercial distributors of home-sized electric wind generators in the United States.

James B. DeKorne was born in Michigan, but spent most of his life in California until 1966, when he became a permanent resident of New Mexico. He has an M.A. in English from San Francisco State College, and has been, among other things, a communicable disease epidemiologist, laboratory technician, college English teacher, and staff photographer for the Museum of New Mexico. At the present time he is living on a one-acre homestead in northern New Mexico with his wife and two children. His main interests are alternative sources of energy in relation to a modern, self-sufficient homestead, and he is trying to form a non-profit foundation devoted to such research at the grass-roots level. He makes his living as a free-lance writer.

Eugene Eccli is co-editor and publisher of *Alternative Sources of Energy* magazine. He is also coordinator of the Environmental Studies Program at the State University of New York at New Paltz where he has taught courses on alternative energy development for several years. His research there has included the design and construction of a low-energy, self-contained solar experimental structure, electrically powered by a wind generator.

Sandra Fulton Eccli authored and illustrated one of the earliest publications pinpointing the problems of centralized control of energy:

1971's *Power Unlimited—or P.U.* She presently co-edits *Alternative Sources of Energy*, and recently edited a book in the field, *Alternative Sources of Energy: Practical Technology and Philosophy for a Decentralized Society.* A former naval officer, she has also been active in the peace, food co-op, feminist and anti-nuclear movements. She is currently serving as technical consultant on "Project Energy-Save," a poor people's program in Roanoke, Virginia.

Robert F. Girvan received his Ph.D. in physics from Iowa State University. After spending two years at the University of Flordia on a postdoctoral appointment in low temperature physics, Bob went on to Boston College to teach physics and solar energy physics. Today, Bob lives with his family in Stanhope, Iowa where he is working on family-sized solutions to the problem of using only renewable energy sources.

Ken Kern is director of Owner-Builder Publications, a non-profit, charitable corporation which is doing research and development of minimum-cost, low-impact technology in the fields of home building and homesteading. Ken, his wife Barbara, and two of their children live part-time in the Sierra foothills where Ken has worked for the past two decades in the fields of land development and surveying, stone masonry, and building design. Ken's background training was in architecture (University of Oregon), and he has written two books, *The Owner-Built Home* and *The Owner-Built Homestead.*

Don Marier received his M.A. in electrical engineering from Illinois Institute of Technology. He taught electronics at Chicago City College, and more recently, worked as an electronics engineer. Don founded *Alternative Sources of Energy* in 1971 and is now co-editor of the magazine.

The New Alchemy Institute is a small, non-profit, international organization comprised of scientists, artists, and humanists who are dedicated to "create a New Science, a science of the earth, shaped not by a few scientists, but by many, many people." The group was created in 1969, for education and research in the fields of ecologically derived forms of energy, agriculture, aquaculture, housing, and landscaping, At the present time, the New Alchemists are experimenting with organic farming, biological insect control, backyard fish farming, wind and solar energy, and methane gas digester systems at their research centers in Woods Hole, Massachusetts; Pescadero, California; and Costa Rica. The Institute publishes a yearly journal and has written several booklets, books, and magazine articles about their work.

Steven M. Ridenour is a research associate at the Institute of Energy Conversion, University of Delaware, where he works with the design, construction, and evaluation of solar collection and storage devices. Steve received his M.A. in environmental engineering from Purdue University. He was a technical consultant for a special project on community utilization of energy and waste recycling for the University of California at Santa Cruz, and has taught courses on alternative sources of energy both at Santa Cruz and at the University of Delaware.

Volunteers in Technical Assistance (VITA) is an international association of more than 8,000 scientists, engineers, businesspeople, and educators from the United States and 50 other countries who volunteer their talent and spare time to help people in developing areas with their technical problems. VITA has published several booklets for constructing tools, power systems, etc.

Ron Weintraub, a former metallurgist, has recently turned his efforts toward experimentation with alternative technologies and energy sources. Ron is a contact person for *Alternative Sources of Energy.*

Sharon and James Whitehurst operate a 100-head dairy farm with their uncle in Benson, Vermont. They first became interested in methane gas production when they heard of the work that Ram Bux Singh was doing with bio-gas plants at his Gobar Gas Research Station in India. It wasn't long after that that Singh himself paid a visit to their farm and helped them build one of the first methane gas digesters in the U.S., which today attracts visitors from all over the country.

INDEX